ST. NORBERT COLLEGE
D0612922

Technology, Man and the Environment

Technology, Man and the Environment

DAVID HAMILTON
Formerly Industrial Editor, *New Scientist*

CHARLES SCRIBNER'S SONS
NEW YORK

1 3 5 7 9 11 13 15 17 19 I/C 20 18 16 14 12 10 8 6 4 2

Printed in Great Britain
Library of Congress Catalog Card Number 73-1595
SBN 684-13368-7

Preface

I have tried here to describe the overall pattern of technology: what technology is, what its effects are on everyday life, how it is changing our world, and the problems it brings. As far as possible I have done this in non-technical terms, but in places it has been necessary to go into some detail to bring out a point more clearly. Those who wish to pursue specific topics will find many good texts to help them. Some are listed in the bibliography.

Throughout the book I have used the metric Système International as far as possible as the preferred system of units, with Imperial equivalents. The figures quoted are generally rounded off and so may not exactly correspond. The unit of the ton has been a particularly awkward one. Three 'tons' are in common use: the long ton of 2240 lb, or 1016 kg; the short ton of 2000 lb, or 907·2 kg; and the metric ton of 1000 kg, or 2204·6 lb. I decided that for my purposes it was not too inaccurate to disregard the differences, and I have therefore chiefly used the long ton without converting the other two where they occurred, which is infrequently.

Many people have helped directly or indirectly in the preparation of the book, and although they are too numerous to mention individually, I owe them all my gratitude. Any faults and discrepancies are, of course, mine and I shall be glad if readers will inform me of them. Nonetheless, if the overall pattern of technology shows through in spite of any blemishes there may be, I shall have achieved my purpose.

Contents

TABLES

Illustrations

PLATES

between pages 48 *and* 49

DIAGRAMS

ACKNOWLEDGEMENTS FOR PLATES

2.1 *United Kingdom Atomic Energy Authority*
2.2 *Taken on a Cambridge Stereoscan: the Plessey Company Ltd, and the Chester Beatty Research Institute*
2.3 *G. W. A. Dummer Esq.*
2.4 *SGS (United Kingdom) Ltd*
2.5 *Mullard Ltd*
3.1 *Fuji Steel*
3.2 *Imperial Chemical Industries Ltd*
4.1 *Baric Computing Services Ltd*
4.2 *International Computers Ltd*
4.3 *Univac Division, Sperry Rand Corporation*
4.4 *Institution of Mechanical Engineers*
5.1 *Hitachi Ltd*
5.2 *Hitachi Ltd*
5.3 *Corrosion and Welding Engineering Ltd (a member of the Constructors John Brown Group)*
5.4 *British Leyland Motor Corporation*
6.1 *The Gas Council*
6.2 *Central Electricity Generating Board*
6.3 *United Kingdom Atomic Energy Authority*

1 · The Anatomy of Technology

The powers that technology has given us are already almost incredible. Technology can sustain men in space and the ocean depths. It can harness the immense surge of the tides and modify the weather. It can blast whole cities to ruins in the blinding flash of an atomic explosion and yet control the same enormous power to yield plentiful supplies of electricity. It can produce on a massive scale substances both to prevent pregnancy and to enable infertile women to bear children—previously a divine prerogative. It has created machines like the heart-lung unit, and artificial, powered limbs that assume functions of the human body for a time. Its crowning achievement, the computer, extends the power of the mind and will have the most pervasive influence upon our world of any machine so far conceived.

What is technology? I should like to define it as the means by which Man extends his power over his surroundings. Our knowledge of the ways of doing this constitute the 'rules' of technology. Strictly speaking, there are many sets of rules for the many different industrial skills and operations by which the process takes place, so there is really no such thing as one all-embracing technology. But it is convenient to lump them all together and talk about modern technology as though it were a single recognizable subject.

Technology is not merely concerned with objects we should normally think of as tools—a saw, a spade, a plough—but with everything else that extends our powers over the environment. Thus the 'tool' may just as well be a bulldozer, a machine-tool, a chemical plant or a production line. We seldom nowadays perform any task, or make anything, without employing a machine somewhere along the line of technological cause and effect. We flick a switch and current leaps out into a light bulb from the long chain of wires and cables, fuses and switchgear, transformers and so on going back to the generators, and if we follow the chain further, it leads us to the fuel needed to run the generators and the materials from which the power station is constructed. We have devices for moving us rapidly around the world—cars, aeroplanes, trains, ships. And others for serving us at work, for helping us run the home and the office without drudgery—e.g. the washing machine and refrigerator, or typewriter and dictating machine. We have devices for enlivening our leisure like television sets and cinema projectors, and for communicating with each other like the telephone and radio. Machines are everywhere. Furthermore, within technology's province we can include our roads and buildings, dams and airports that envelop ever greater portions of the natural environment beneath a man-made veneer. We can include everything, in fact, with which we support our life at its complex and sophisticated level.

Different people understandably see technology in different guises. For the factory worker it has overtones of 'automation', and it gives him visions of new pieces of machinery—and possible redundancy; in any case it is something to be viewed with reserve, if not apprehension. The businessman may think in terms of a computer taking over routine payroll calculations previously carried out by girls with accounting machines, and some troublesome matters of organization and communications that need to be solved before his firm can reap the expected rewards. To the man in the street it may well suggest space exploration and landings on the moon. The scientist may perhaps think of technology in terms of finely-engineered apparatus in his laboratory. The housewife is more likely to think of it—if she thinks of it at all—in relation to her washing machine, plastics kitchenware, synthetic-fibre clothes or transistor radio.

In the course of the book we shall look at technology from several of these viewpoints: technology in the materials we use and the methods we employ to shape them; technology in the operations of the 'high technology' industries such as aerospace, computers, telecommunications and chemicals; technology in the fight against Man's age-old enemies—thirst, hunger, disease. We shall see its impact in the home and the work-place; and its effect on people, on government and on the world.

Technology and the Environment

First, the basic question of the effect of technology on the natural world—our environment. Considered in its broadest terms, the action of technology is to take materials from the environment at one point, make use of them with the help of some form of energy, and then return them to the environment at another point: dust to dust, ashes to ashes. Let us examine this in a little more detail.

Sometimes the materials can be extracted and used as they are. The energy required is low, and a man's strength can easily supply it. Clay, stone and wood are often used in this way. They were good enough for our early ancestors, who originated technology when they used stones and bits of wood for killing and skinning animals—perhaps for fighting among themselves—and made pots for keeping their grains in. Almost certainly this capacity for using tools helped early Man to survive. The earliest known man-made tools, some two million years old, were found in Africa. It appears that for many thousands of years around that time Man co-existed with animals that were physically his equal. These crude tools must have given him some advantage over them in the exploitation of his surroundings.

Our powers now are infinitely more extensive, and our 'tools' correspondingly more complex. In most cases we employ some form of chemical and physical treatment to refine the natural materials, or transform them into new ones that we can use. Steel, our most widely used material, is a prime example. Iron ore, a brownish earth, must first be reduced to iron in the white heat of the blast furnace. Impurities are roasted out of it to yield steel, and the crude steel ingots are shaped by rolling, forging, and other processes into bars, tubes, rods, sheets and so on. Some of these are used directly but most are machined further to make components that are incorporated into machines and many other man-made objects.

Plastics undergo a similar preparation (see page 61). Crude oil is carried by tanker from oilfield to refinery and there distilled and treated to make various raw materials for chemical synthesis like ethylene and acetylene. One or more of these materials, and probably other compounds, are reacted over a catalyst to produce a plastic. The plastics grains are heated and moulded under pressure to make the familiar plastics articles.

These materials—steel, plastics and all the rest—constitute our tools, machines and constructions. Sooner or later, the used materials are discarded. They are not 'thrown away', but transferred to another part of the environment, whether it is land, sea or air. Matter may be converted into energy—as when we burn a fuel—that is absorbed somewhere in the environment. The only irretrievable loss occurs when energy is radiated into space, although often a material once used is not recoverable in its original or converted form. Plastics are often dumped in the ground, or burned, when they are converted into heat and gases that stream into the air. Steel parts that rust away also return to the earth. Other parts, recirculated as scrap, go through the production cycle again—bypassing the blast furnace—losing a little of themselves to the environment each time as slag or in the form of combustion gases, or waste liquors. For a short while, we have the materials for our use. Eventually, however, the environment claims them back, just as it claims us.

Technology helps Man in his eternal struggle for survival in a hostile world. In order to obtain food, warmth and shelter, Man has to arrange his environment to suit his needs. These needs are usually not those of the other inhabitants—animal and plant—of the planet. His modification of the environment brings him into conflict with Nature. He works towards order, Nature towards disorder. Nature continually tries to bring his works to ruin. In this continual struggle, both sides may suffer. For most of his history Man has been losing. Today his machines have given him the upper hand: but this power over the environment can be a dangerous thing, and some of the changes he is making—sometimes unwittingly—are positively harmful. Without doubt, the environment has been changing ever since it was formed, and Man has merely added to the changes wrought by natural forces. What matters, though, is the scale, pace, and sometimes irreversible nature of these modifications. While technological power was limited, the scale and pace were, too. For thousands of years Man's activities amounted to little more than tending fields and flocks, cutting trees for timber or fuel, working metals and other materials on a small scale, and living in small communities in essentially rural surroundings. That was the lot of the great bulk of the world's population until recent times. But now we are drastically affecting our environment: indeed, technology has given us the power to alter the whole world. If the hydrogen bomb is probably the most potent of its devices, there are many more subtle but no less hazardous influences at work. Foremost are the pollution of air and water and the disposal of tremendous quantities of solid wastes as mass production and mass consumption spread throughout the world.

The Pace of Change

Man's modification of the environment is taking place much more rapidly than in the past. The pace of change is breathtaking. In 3000 years ancient Egypt hardly altered: the world into which a man was born was the one in which he died. Yet there are now people living who were born before cars, aeroplanes, radio, television, plastics, drugs, computers and gramophone records became commonplace, perhaps even before they were invented. In general, the speed with which new discoveries are turned into useful products is increasing as time goes by. This is shown in Table 1.1.

Table 1.1

Time Taken for Exploitation of Discoveries

Approximate date of discovery		*Time taken from discovery to exploitation*
1831	electric motor	50 years
1861	telephone	15
1887	radio	7
1925	radar	9
1925	television	9
1931	atomic reactor	11
1938	atomic bomb	7
1948	transistor	5
1958	integrated circuit	3

The rate of change is itself accelerating. New technologies spring up to oust old ones. New industries arise to attack senile ones. New products are launched on the market to enjoy a brief life before they are vanquished by newer ones. Industries themselves are growing more rapidly in this century than ever before. The electrical industry, for example, grew quite slowly out of the discoveries of Michael Faraday and other pioneers in the nineteenth century, but the nuclear power industry has leaped forward from its starting point in the scientific discoveries of the 1930s. The electronic computer industry has grown into a vast undertaking in the twenty-five years since the first rudimentary machines were constructed. Although there are exceptions to this general rule, overall we are far quicker to exploit discoveries than we were in any previous age.

Information, Technology's New Base

Together with its pace and scope, the nature of technological activity is changing, too. We are in the midst of an industrial revolution no less fundamental than the one in which modern technology was born some two hundred years ago. The present revolution involves the communication and use of information, just as the first one involved the transport and use of energy supplies. It is coming about because technology is becoming more information-centred. The technology of the past depended on human experience and skill, and often intuition. Modern technology is coming to depend almost exclusively on science.

There has been a growing interaction between technology and science since the great awakening of scientific curiosity in the seventeenth century. What used to be done by rule of thumb and often plain luck is today governed by rules of chemistry, materials science, quantum theory, and the like. Machines and processes are described by engineering drawings, patents, test data, computer programs, by articles in books and pamphlets, journals and papers. The rules of technology have become well enough defined for them to be put on paper or into other media and passed from one man to the next. That is to say, technology can be traded in terms of information as well as machines. The information side assumes greater importance as technology becomes more advanced and more susceptible to mathematical formulation.

In this trade, information in general flows from nations possessing more useful or profounder knowledge to those less fortunate. By this I mean that information flows from an area of high technology to one of low technology, like water running downhill. The industrialized western nations together with Japan mostly have a higher level of technology than Asia, Latin America and Africa. Information flows from the industrialized nations across the technological gap to the rest of the world, and raw materials and other goods flow back to pay for it. Among the industrial nations themselves there are different levels of technology. A nation like the United States, with its high technology level, collects a large amount in 'knowledge payments' even from other industrial nations. The trade in information—measured by such things as the money spent to license another country's inventions—is a good indicator of the relative levels of technology in different areas. Table 1.2 sets out the

Payments for Patents, Licences and Technological Know-How
($ millions)

Payments to → / Payments by ↓	UNITED STATES	UNITED KINGDOM	GERMANY	FRANCE	ITALY	JAPAN	SWITZERLAND	CANADA	NETHERLANDS	BELGIUM	SWEDEN	DENMARK	*Total payments to 12 countries*	*% of total*
UNITED STATES 1964	*	21·0	10·8	11·7	2·3	5·0	n.a.	37·0	n.a.	n.a.	n.a.	n.a.	87·8	12·0
UNITED KINGDOM 1964	81·8	*	3·4	11·7	2·2	0·3	8·4	n.a.	—	n.a.	0·8	—	108·6	14·8
GERMANY 1964	65·3	17·2	*	5·3	1·7	0·2	42·8	2·0	11·2	1·3	2·4	1·5	150·9	20·6
FRANCE 1963	59·7	11·9	7·5	*	2·0	—	27·0	0·7	4·6	3·8	1·2	1·0	119·4	16·3
ITALY 1963	57·2	15·0	15·2	14·6	*	0·4	22·3	0·7	5·6	2·1	2·0	0·4	135·5	18·5
JAPAN 1963	84·7	11·0	12·5	3·0	1·7	*	9·5	0·8	4·6	0·7	0·7	1·0	130·2	17·8
Total receipts from the above 6 countries	348·7	76·1	49·4	46·3	9·9	5·9	110·0	41·2	26·0	7·9	7·1	3·9	732·4	
% →	47·6	10·4	6·7	6·3	1·4	0·8	15·0	5·6	3·5	1·1	1·0	0·5	100·0	100·0
Total receipts from all countries	550·0	121·5	61·6	47·2	32·5	5·9	—	—	—	—	—	—	—	—
As % of total receipts of 6 countries	67·2	14·8	7·5	5·8	4·0	0·7	—	—	—	—	—	—	—	—

n.a.—not available

balance of trade in information between major industrial countries.

This means that a supplementary cycle concerned with the transfer of information on the rules of technology and the question of how they can be used to advantage is being superimposed on the basic material cycle of technology I mentioned earlier. The two cycles have certain similarities. In the one, materials are extracted, used and finally returned to the environment. In the other, information is originated, used and then put into store or destroyed. The chief difference is that information disappears when the medium on which it is stored—a sheet of paper, say, or a photographic film, a computer memory, or the human brain—is destroyed.

It is estimated that by the late 1970s half the wealth of the United States will be provided by the information industries (see page 151). Britain and other countries are treading the same path. Our children and grandchildren will many of them earn their living in these new industries.

Losses and New Opportunities

There are plenty of new opportunities in the newer industries, the information industries included. Looking back over the past half-century, we can see the origins of some of these opportunities in new technological devices. The major success story belongs, of course, to the motor car, the aeroplane and the internal-combustion engine. The tremendous growth of the motor-vehicle industry has created new jobs in its own factories and those of its associated producers—the making of steel plate, forgings, rubber tyres, petrol and diesel oil, lights, glass, electrical batteries and generators, and many more items.

The continuing demand for more fuel has led to the construction of highly automated oil refining plants, oil and gas wells, mammoth tankers, and advanced drilling and prospecting methods. The past twenty-five years have seen the emergence of nuclear power as a new source of energy with all its potential benefits and its potential hazards. The innovatory industries of electronics, aerospace and chemicals derive their very existence from rapid technological advances. Even the leisure industries provide relaxation for the workaday man through increasingly sophisticated technology: just take the examples of gramophone records and record players, broadcasting or the film industry. As a general rule, these newer, innovating industries provide work that is physically less demanding than the heavier, more basic industries.

Because of increasing mechanization in manufacturing and in the basic trades like agriculture, fishing, and mining, fewer people with more machines can produce all the material goods required—though more people are needed to handle the paperwork involved in producing, selling, distributing these goods and looking after the production workers. That is, they handle information instead of materials. More people find their way into banking and insurance, government service, law, accountancy, the professions and public services. Each year nearly a quarter of all the young people leaving school in Britain find jobs in offices. If in the past forty years the labour force in Britain's manufacturing industry has increased by about a third, the office population has increased twelve times. The United States is the first nation in history in which less than half the employed population is involved in the production of food, clothing, housing and other tangible goods. The economy is becoming more and more concerned with services, research and development, education and activities made possible by larger amounts of leisure time. Three of its technologically orientated industries, virtually non-existent in 1945—television, jet aircraft, and the computer—were by 1965 contributing more than $13,000 million to the gross national product of the United States, and had created an estimated 900,000 jobs. More recently, it has been estimated that the 'convenience' business of producing dishwashers, waste-disposers and self-cleaning ovens has created 50,000 jobs in the appliance industry in the United States within ten years, and at least as many in related business and industry.

It is a sign of a maturing economy that the secondary occupations—like the information industries—become more important than the primary ones. Each nation as it becomes more heavily industrialized, more developed in its technology and more reliant on information for its industry, reaches this stage. The United States, Britain and others have already reached it. Some of the developing countries have a very long way to go.

While the new industries enjoy rapid growth, some senile ones are flagging. Unless they can adopt new technology, they are doomed. The coal industry in Europe and North America, for instance, has suffered dreadful contractions in recent years. In Britain pits were being closed at the rate of about forty a year at the beginning of the

1970s and each year about 30,000 men were leaving the mines. Other traditional bastions of Britain's might have been eroded as well. Shipbuilding and marine engineering have felt the keen draught of competition from Japanese, German and Scandinavian yards. Employment in quarrying and fishing has been falling steadily. Some 230,000 workers have left agriculture in the last ten years. Between 1961 and 1966 the British industries that lost the greatest numbers of workers were shipbuilding and marine engineering (they lost 25 per cent of their workers), mining and quarrying (21 per cent), and textiles (9 per cent).

Diagram 1.1 shows how employment in Britain has changed over the past half century, with people transferring from the primary production industries to the secondary, service industries.

There will, however, be other jobs available as long as industry is healthy and accepts new technologies, devices and processes, and employs new information skilfully in management. The individual

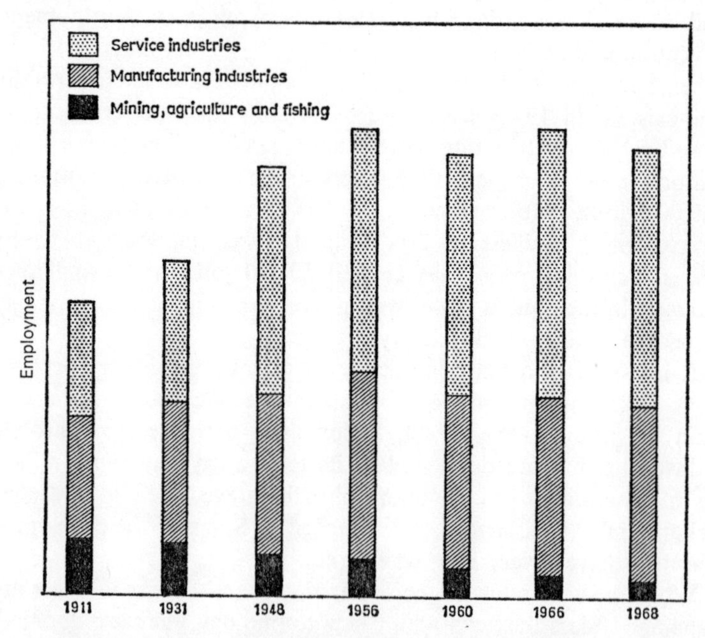

Diagram 1.1 Over the last 50 years employment in the primary industries such as agriculture has fallen and more people have found jobs in the service industries (after K. S. Reader, *The Modern British Economy in Historical Perspective*, Longman, London, 1969)

may have to pay the price of this technological evolution: he may have to move from one of the older, static industries to one of the newer ones to find a job. Because of rapid changes in methods and machinery, workpeople will have to accept that the skills they start out with may not sustain them through to the end. Skills quickly become outdated.

The Value of Technology

Without a doubt, technology has played a major part in raising the world's living standards. By providing more machinery to aid the worker, it has enabled him to produce more with less effort, whether it is in the factory or the fields. According to one estimate, technological advance in the United States accounted for 90 per cent of the rise in output per man-hour between 1871 and the mid 1950s. A similar sort of conclusion could almost certainly be reached for other major industrial powers.

To be commercially successful, the modern industrial state needs four main things: innovation in technology, including investment in new machines and processes; wide-ranging education; cheap, plentiful energy supplies; and rapid communications. Let us look at each in turn.

Innovation is generally understood as the whole operation of developing a new device from discovery to marketing. It starts with research. A nation (or company) can try to induce innovation in certain areas of technology by spending more on research in those areas. In industrial nations, government commonly pays for 50 per cent or more of the nation's research and development, and government can therefore influence the rate at which technology progresses in certain fields merely by adjusting the balance of funds given to the different fields. However, there is a snag. Research and development do not necessarily lead to innovation and worthwhile new products. If they did, Britain would be a good deal wealthier. Like other European countries, though unlike the United States, Britain is better at fundamental research than translating the results into useful products and selling them. A classic case was the aniline dye industry. The basic discoveries were made in Britain in the late 1850s and early 1860s but they were soon exploited on a massive scale in Germany. The more recent examples of radar and computers show that the situation has changed painfully little except

Table 1.3

Deployment of Scientific and Technical Talents

	Country	*Investment % GNP*	*GNP $/head*	*% in agri-culture*	*Standard of living* Use of electricity kWh/head/year	*Number/1000 people* Cars	Phones	TV sets
	USA	16·6	4,380	5·0	6,532	414	540	392
Sizeable industrial countries	UK	18·2	1,850	3·1	3,481	196	218	280
	France	24·9	2,530	15·8	2,216	240	141	185
	Germany	23·1	2,200	10·2	3,088	194	172	248
	Italy	19·4	1,390	22·5	1,764	136	132	144
	Japan	34·0	1,400	19·7	2,345	48	107	206
Communist countries	USSR	?	?1,280	27	2,500	5	40	?
	China	?	?150	?80	?60	?	?	?
Smaller industrial countries	Austria	23·6	1,550	20·0	2,538	132	159	134
	Belgium	21·1	2,160	5·6	2,664	160	184	185
	Canada	23·3	3,010	8·6	7,780	282	408	284
	Netherlands	26·5	1,980	7·9	2,333	157	203	200
	Norway	26·7	2,360	15·5	13,354	156	255	179
	Sweden	23·6	3,230	9·3	6,432	246	489	289
Developing countries	Greece	26·0	860	50·1	715	18	67	—
	Ireland	19·1	1,070	29·6	1,432	110	87	111
	Portugal	19·2	530	32·3	590	30	65	29
	Spain	20·9	770	29·4	1,109	50	113	90
	Turkey	19·4	350	72·7	183	4	9	—

As far as possible, these figures are comparable and relate to 1968. But they should be treated with caution; they are intended as approximate indicators of relative positions, not as absolute values. The entry against Belgium also includes that for Luxembourg.

in the Industrial and Developing Countries

	Country	Gross national spending on R & D: $ million	Gross national spending on R & D: as % GNP	Scientists, engineers and technicians on R & D, full-time equivalent	Qualified scientists and engineers per 10,000 population	R & D expenditure $/person
	USA	21,075	3·4	696,500	35·8	110·5
Sizeable industrial countries	UK	2,160	2·3	159,540	29·4	39·8
	France	1,299	1·6	85,430	17·9	27·1
	Germany	1,436	1·4	105,010	18·0	24·6
	Italy	291	0·6	30,280	6·0	5·7
	Japan	892	1·4	187,080	19·5	9·3
Communist countries	USSR	8,000	2·4	550,000	30	?35
	China	700	1·0	?65,000	?4	1
Smaller industrial countries	Austria	23	0·3	3,220	4·5	3·2
	Belgium	137	1·0	15,600	16·8	14·7
	Canada	425	1·1	23,850	12·6	22·5
	Netherlands	330	1·9	31,310	25·8	27·2
	Norway	42	0·7	3,820	10·4	11·5
	Sweden	257	1·5	16,530	21·6	33·5
Developing countries	Greece	8	0·2	1,260	1·5	0·9
	Ireland	10	0·5	1,670	5·9	3·5
	Portugal	9	0·2	2,230	2·4	1·0
	Spain	31	0·2	6,480	1·5	0·9
	Turkey	27				

As far as possible, these figures are comparable and relate to 1963/64, except those for the Soviet Union and China, which are for 1965, with some allowance for recent trends.

that these days it is the United States that does the exploiting. At present, Britain spends about £1,000 million annually on research and development. At about 2·7 per cent of the gross national product, it is a higher fraction than that for any country apart from the United States and the Soviet Union. Yet Britain has a lower rate of economic growth than most other industrial countries. Japan, on the other hand, spends relatively little on research and development—although the funding is being increased—but has achieved astonishing rates of growth partly through exploiting western technology: no less than 14 per cent in 1968, for example.

Innovation needs considerable investment not only in research and development but also in engineering equipment, jigs and tools, design drawings, market research and so on. It cannot succeed unless funds are forthcoming either from governments or from private investors. As a very general rule, spending on research and development is not more than about one-tenth of the total investment needed to bring an invention to the marketing stage. Thus, if the research and development of a new machine has consumed £1 million, another £10 million will be required to get it into production and on to the market. Investment in technology accounts for the majority of this, leaving aside such matters as advertising, distribution and servicing.

Some countries find it easier than others to provide such investment, which is necessary for converting the discoveries of science into the products and processes of technology.

The next table—Table 1.3—relates these different factors, showing how various countries, industrial and developing, deploy their scientific and technical talents, how much they invest of their national wealth, and what rewards they get from their efforts. The figures are taken from the most recent international comparison made by the Organisation for Economic Cooperation and Development (OECD) in 1963–64, supplemented with some more recent ones from OECD and other sources. It shows the United States to be the most massive spender on research and development, a country with the highest standard of living in the world and a very small proportion of its people working on the land. Other industrial countries cannot emulate either quite that pitch of research and development or such a standard of living, though some do come pretty close. What does stand out is the great gap between the rich, industrial countries and the poor, developing ones. The developing countries mostly have a large fraction of their working population on the land and a low average income, a low standard of living and hence a low total for spending on research. It is a vicious circle. The nations that can afford to do research and development become technologically more capable and their products more competitive; those that cannot afford it—because there are more urgent claims on their money—cannot advance. So the gap between the two groups widens. The actual totals shown in the table have changed since the dates given, but the relative position is still substantially the same. Japan, the Soviet Union and other nations have been able to achieve rapid economic growth mainly through transferring men from the low productivity of peasant agriculture to the high productivity of manufacturing industry. It is a process that Britain, like other industrial countries, has more or less completed.

The industrialized world spends a good deal of money on research and development, which in turn gives new life to technology. On comparing the totals for spending on research and development in different areas, however, we find that spending is relatively low in the areas most relevant to mankind's basic needs such as social sciences (apart from education), town planning and land use, building, food production and agriculture. It is relatively high for items relating to war and space exploration: weapons, aerospace, computers, electronics, telecommunications, nuclear power. The next table—Table 1.4—demonstrates this as it applies to certain industries in Britain, although the situation is typical of most industrial countries.

The odd and worrying imbalance implies that the human race is more interested in destruction and the rest of the universe than in sustaining life on its own planet. The pattern of inquiry in research has long been distorted by this leaning towards the physical sciences—which are those that have been employed for military ends—and away from the 'human' sciences. It is a pattern that developing countries tend to imitate, to their detriment. India, for example, for some time spent about 40 per cent of its research budget on nuclear physics and only 8 per cent on agriculture. As a result Indian nuclear physicists are among the finest in the world, but Indian agriculture was neglected.

Because of the immense funds and the vast numbers of scientists and engineers that it was able to apply to research and development, the United States was able to build up a commanding position in the

1940s and early 1950s in many spheres of electronics, computers, aerospace and nuclear power technologies.

However, both there and in other industrial countries the bias is now changing owing to doubts about the aims and activities of science and technology serving military or quasi-military ends. If space rockets and computers glitter with the stardust of the

Table 1.4

Research and Development in Manufacturing Industry

Industry	*UK research and development spending as per cent of output in 1964–65*	*Output as per cent of that of all manufacturing industry* (1) UK 1963	(2) US 1965
Aircraft	39	3·1	3·9
Electronics	14	3·7	4·8
Petroleum products	13	0·7	1·9
Scientific instruments, clocks	6	1·5	2·4
Chemicals	5	9·3	9·1
Motor vehicles	4	7·1	7·6
Electrical engineering	4	5·6	4·5
Rubber	2	1·7	2·6
Mechanical engineering	2	14·5	10·6
Glass, pottery, bricks	2	4·1	3·6
Primary metals	2	7·9	8·8
Ships and marine engineering	1	2·0	0·7
Textiles	1	8·0	3·5
Food, drink, tobacco	1	12·2	11·8
Timber, furniture	less than 0·5	2·7	3·6
Paper, printing, publishing	less than 0·5	8·5	9·4
Clothing, footwear, leather	less than 0·5	4·5	5·1

Source: *Technological Innovation in Britain*, HM Stationery Office, London, 1968.

technological age, the Americans' eyes are being forcibly turned from the stars to the earth; from rockets and satellites to the problems of their own society—to topics like air and water pollution, education and welfare, the plight of poor people, food supplies and population control. Further industrial nations are involved in the same reappraisal, although the United States, technologically the most advanced, has been the first to feel the need for a new set of values. It has felt it keenly.

Like innovation, *education* is one of the most important influences working for technological progress. Like innovation, too, it cannot easily be adjusted or turned up or down to produce the sort of results that are wanted by the state. In this field, which is basically one of human interactions, results cannot be guaranteed purely by expenditure. In the educational process information flows from people who possess profounder or more useful knowledge to those who are less fortunate, just as it flows between nations. Nonetheless, without expenditure there will be no results whether in terms of teachers, schools, or educated pupils. In most countries spending on education is listed as a priority and it has risen sharply since the last war.

The aim may be to produce people knowledgeable in all kinds of subjects—from history to maths, languages to science, music to metalwork—and continue their education through an integrated system for the whole of their lives. Unfortunately, this is unrealistic. The boundaries of scientific knowledge alone are being pushed back at such a rate, and the derived technologies are so complex, that even in a three-year college course the student can do no more than glimpse a small part of the whole vista. So there has to be very much of a compromise.

That is where the discussions and the disagreements start. All countries, it can be assumed, are anxious to improve their educational systems, if for no other reason than that they can run their affairs better with educated people than illiterates. But there is more to it than that. Because of the increasing dependence of life on technology, and of technology on science and knowledge, a rising level of skill is demanded of the bulk of the population. They will need to know how to cope with machines they have not encountered before, and how to find, understand and use information given to them in many forms.

Different nations and different authorities within each nation have radically different ideas about how people ought to be educated in these fields. Some prefer specialized courses, others general ones. No course can be both broad and deep; there is not the time for it. Either it is narrow and deep, or broad and shallow, and there is endless debate about the extent of the compromise. The balance between arts and sciences also comes in for a great deal of discussion. An industrial country realizing how important science and technology are for its economic growth wishes to bring up as many of its people as possible to be informed in these subjects. But of course not every student wants to study them. Nor may there be a suitable job waiting for him when he leaves his school or college. Besides that, in the past, education in science and technology has concentrated on producing people who are trained to solve technical problems rationally without thinking of the cost or consequences of their technical solutions. That has brought disrepute upon science and technology, discouraging many able young people from pursuing them into their working lives.

Energy, the third of the prerequisites for the modern industrial state, falls into a different category from the two we have discussed. It is a commodity that can be measured, bought and sold, and brings about predictable effects.

Man needs fuel to give him warmth, light, power. Civilization began in the world's equable climates—around the 21°C isotherm—and moved into cooler areas as he learned to clothe himself and use fire. Modern technology permits civilization to be established virtually anywhere through the exploitation of new energy sources and conversion devices. Historically, it was not until the power of muscle, wind and water were augmented by the power of steam, and later the internal combustion engine, that technology could advance with anything but faltering steps. It is true that some of mankind's great works were raised by men sweating against natural forces with only rollers and levers to help them. The pyramids of Egypt and the stone megaliths of Stonehenge in England and Carnac in France were raised some time between 3000 and 1000 BC using these methods. But such achievements were only feasible with the concerted efforts of thousands of slaves. The power commanded by one man was, by today's standards, pitifully inadequate.

Again, it is true that there were significant inventions before there was much power to make them work. Between A.D. 200 and 1600, for example, Man added the mariner's compass, clocks, windmills, and the processes of printing and distillation to his store of technological devices and skills. The lathe, the screw-press, and the pump were among the devices known to the Chinese before A.D. 400.

Significant though these inventions were, they lacked one essential attribute—substantial amounts of power. The widespread use of machines, characteristic of modern technology, did not become feasible until more potent sources were harnessed to drive them. The first Industrial Revolution was marked by a great increase in energy consumption, and this quantity is still used as a yardstick in assessing a country's industrialization—and hence its adoption of technology.

The first such power source, the steam engine, was initially a lamentably inefficient apparatus, converting a derisory amount of the heat energy of the fuel into useful energy. As first devised in the West Country in England in the seventeenth century, it was employed to pump water from mines. If James Watt made it a practical source of power, contrary to usual belief he did not invent it. The Watt engine was set to work to turn Britain's lathes and drills, spinning and weaving machines, and all the other devices of the new-found technology. The machines were grouped around the steam engines installed in factories, which gradually replaced the cottage or the forge as the location of industry. The transfer of industry from cottage to factory accelerated the migration of people from the farm to the town, and ushered in the change from agricultural to industrial economies that most western nations, and a few eastern ones, have now largely accomplished.

The other main trend has been the adoption of nuclear power and the growth of the industry making nuclear equipment. Although still in its infancy as regards the amount of electricity generated by means of nuclear reactors, the nuclear power industry holds out the distant vision of cheap, plentiful power and will have a firm foothold in the industrial structure of the future. All the western industrial nations are pursuing their own lines of development here, and all are trying to recoup some of the costs by selling their reactors and nuclear fuels to the other industrial countries and the developing world.

Solid fuels, liquid fuels and now atomic fuels, together with various power conversion devices (ranging from the internal combustion engine, steam and gas turbines, the electric motor, and the fuel cell to the generator and the battery), have helped accelerate technological progress and support the vast edifice of technology. Nowadays the price and availability of energy play a major part in running existing industries and starting new ones. Canada and Norway are blessed with ample, cheap hydro-electricity and have become prime operators in smelting aluminium. Britain's plentiful and accessible coal reserves provided both fuel directly and coke for the iron-masters' blast furnaces, and so acted as catalysts for the industrial revolution.

The fourth, and last, important attribute of the modern state is good *communications*. A state with good communications can rapidly transfer information on technology—and a thousand and one other topics—between various places within its borders, and carry on a ready conversation with those outside, reacting to changes swiftly. In future the trade in information will have just as much effect on the economic well-being of a state as innovation, education and energy because technology, which is largely responsible for raising living standards, leans heavily on the transmission of information. Communications systems of all kinds are having to be strengthened to take the load of quantities of data, conversations, radio and television broadcasts, and written and printed messages.

Communications in its wider sense means more than this. It also covers the carriage and handling of goods and of people. The network by which information is transmitted from person to person (along a telephone line, say) has its counterpart in the network of roads, sea lanes and aerial highways. We cannot press the analogy too far, since there is no 'carrier' in the telephone wire that corresponds to the lorry or car on the road, or the ship or aircraft; in certain ways, however, the process is the same. Items and components must be transported from factory to warehouse and from warehouse to retailer or user, etc. The transport of people poses difficult problems, but if more conversations could be conducted through video-telephones instead of face to face it would eliminate a great deal of pointless travelling, and reduce the congestion of human transport systems.

Already technology has shrunk the world into a 'global village'. Not many years ago, it took a fast ship five days to cross the Atlantic. Now we can fly it in five hours or less. We can watch events on the other side of the world, as they happen, by television signals relayed to us from satellites hovering 36,000 km (22,300 miles) above the earth. Without rapid communications an organization becomes sluggish and indecisive. Technology has given us the means to make the world and its organization as responsive and as vital as we like.

The Developing World

If technology brings affluence, this has not been distributed equally about the world. We saw that there is a gap in the level of technology between different parts of the world and, as a result, the disparity in living standards between the rich, industrial nations that have discovered how to turn technology to their advantage and the poor, developing nations that have not, is growing wider. Aid from the industrial nations is stagnant, and the developing countries face grave problems in trying to earn enough from the sales of their raw materials—often all they have to offer—to pay for the manufactured goods and advanced technology that they want in return. Some of the developing nations have been able to redress the balance by selling fuel or raw materials for which the industrial world is willing to pay high prices, notably oil and nuclear fuel ores. Countries with reserves of nuclear fuel ores—chiefly uranium and thorium (which can be converted into fuel by irradiation in a reactor)—possess valuable assets that only forty years ago would have been unrecognized. Until the last war, for instance, uranium was valued merely for the yellow colouring properties of its compounds. The oil lands of the Middle East, Africa and South America are reaping rich harvests from their natural bounty. But overall, the developing countries—those that need technology most urgently—are the least able to afford it, and can only sell agricultural or other commodities (coffee, sugar, cocoa, cotton) in payment at prices that are liable to fluctuate wildly according to the way Nature treats the crop. Many organizations, headed by the United Nations and extending through a gamut of government, official and other agencies, are trying to help the developing countries direct through education, agricultural, medical and other programmes. The bulk of the UN's effort, for example, is aimed at tackling the problems of the developing countries—the Third World, as they have been collectively labelled—and closing the gulf between them and the industrialized nations.

In spite of their efforts, the rich nations of north-west Europe and those elsewhere in the temperate zones that were settled and organized by people of the same stock still produce and consume about two-thirds of the world's goods, although they claim but one-third of the world's population. Moreover, 90 per cent or more of the world's industrial output is still concentrated in areas inhabited by people of European origin.

One of the most intractable problems of the developing countries is coping with the great influx of people to the cities that occurs in the course of industrialization. The nations that are now industrialized went through the same experience, but a hundred years ago or more. They had plenty of time. The developing countries today, in attempting to introduce twentieth-century technology rapidly into their age-old agricultural economies, face severe upheavals and misery. The transfer of people from the country to the town—industrialization coupled with urbanization—inevitably creates stresses as building programmes, public health, education services,

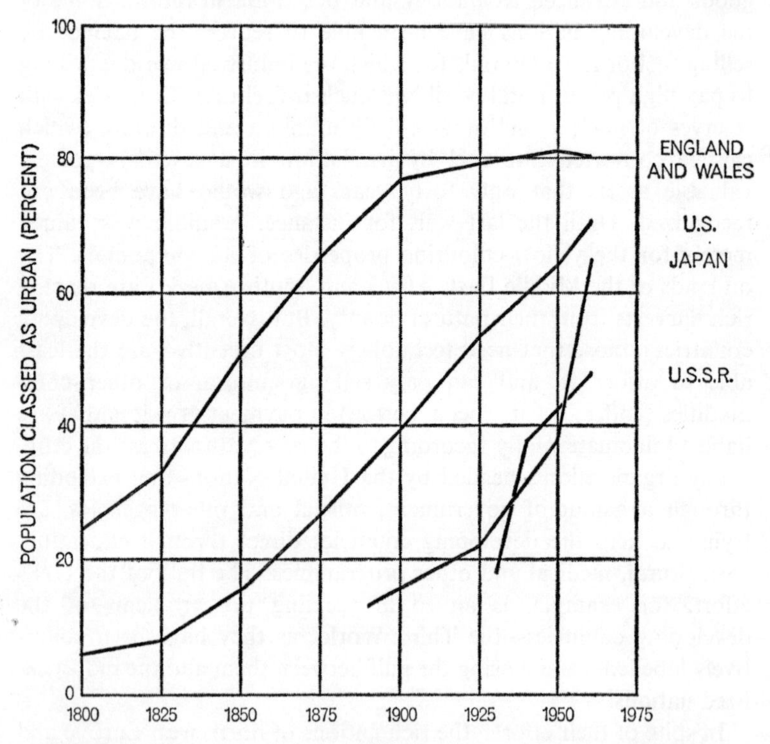

Diagram 1.2 The currently industrialized nations underwent a process of urbanization typified by the curves shown here for four of them. It was closely related to their economic development.

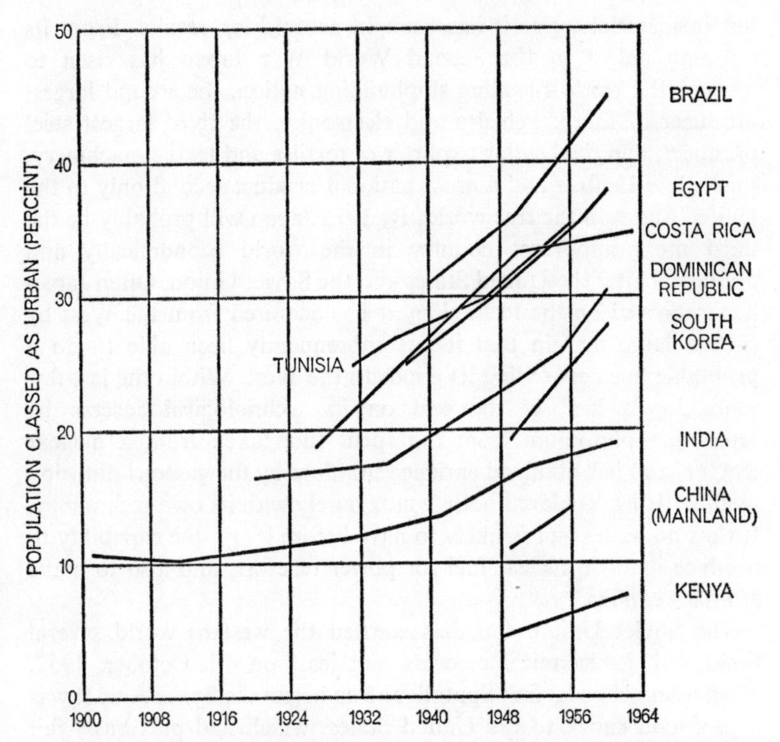

Diagram 1.3 The developing countries of today started to become urbanized much later than the industrialized nations, and their urbanization is coming about at a slightly quicker pace.
(Acknowledgement for Diagrams 1.2 and 1.3: from "The urbanization of the human population", by Kingsley Davis. Copyright © 1965 by Scientific American, Inc. All rights reserved)

food distribution and other services try to keep pace with the numbers crowding in, looking for work. The speed with which urbanization is taking place in the developing nations is apparent in the two diagrams above, which compare the process as it occurred in western Europe, the United States and the Third World.

As the result of better communications, overseas investment by large companies, deliberate stimulation through aid programmes, and other methods, technological skills are diffusing throughout the world, more quickly in some areas than others. Japan presents a fascinating picture of western-style technology impressed upon an oriental culture and social structure. This creates tensions naturally, but in material terms it has brought astonishing results. From its crushing defeat in the Second World War Japan has risen to become the world's leading shipbuilding nation, the second largest producer of motor vehicles and electronics, the third largest steel producer, and the leading exporter of textiles and textile machinery. In the late 1960s it had a gross national product second only to the United States in the free world. By 1975 Japan will probably be the third most important country in the world economically and politically after the United States and the Soviet Union. Often Japan has improved on the technology it has acquired from the West by such a large margin that it has subsequently been able to do a profitable business selling its goods to the West. Within the last few years Japan has set the seal on its technological success by separating plutonium from the spent fuel taken from a nuclear reactor, and has obtained enriched uranium by the gaseous diffusion method. It has achieved both results purely with its own technology. It thus possesses—or is likely to have before long—the capability to produce its own nuclear fuel for power reactors, and also to make atomic weapons.

The Soviet Union, too, has startled the western world several times with its technical prowess, not least on 4th October, 1957, when it launched its first Sputnik and so began the Space Age. It was a profound shock to the United States, which had previously felt securely in the lead in technology. From then on the race between the two superpowers has diverted the course of technology into projects whose value is at the least questionable, and which have certainly consumed enormous amounts of money, time and man-power, and materials.

Since that time, France and China have exploded nuclear weapons, and have developed missiles to carry them. Israel and other states are almost certainly developing nuclear weapons, following the example of the industrial powers. Unhappily, no one has yet found a way to prevent advancing technology from being harnessed for destruction. In order to illustrate how recently we have acquired these technological powers, let us look at things this way. Man's evolution, it is usually accepted, has taken about two million years from the first terrestrial appearance of man-like, tool-making apes. If we liken that to a single year of our time, then the earliest civilization on earth began after 5 p.m. on 30th December at the end of that year. That is, if one year accounts for the evolution of Man, then one day and seven hours account for all his twenty-five civilizations, starting with the dawn of ancient Egyptian civilization before 4000 B.C.

On the same scale, modern technology dating from the British Industrial Revolution has been with the human race for about an hour and twenty minutes; practical long-range communication, beginning with the telegraph, for about forty minutes; and modern electronics, based on the transistor and other solid-state devices, for no more than eight minutes. The first heart transplant was performed a mere twenty seconds or so ago. The average person lives about twenty minutes.

Technology has given Man almost unbelievable powers in no more than the twinkling of an eye compared with the time it has taken him to evolve. Assuming that his race is not wiped out by a cosmic cataclysm, he will have to act wisely if he wants to ensure that human life continues on the earth or elsewhere for anything like the span he has already enjoyed.

Technology can achieve practically anything today if we spend enough on it. It gives Man unprecedented powers over his environment and over himself. For the first time in history, all mankind, and not just a tiny, privileged majority, can aspire to the benefits brought by wise husbandry of the environment—more nutritious and more varied foods, better houses and urban conditions, better facilities for medical care, education and welfare, and other amenities. There are few technological barriers left in the way. Virtually everything is possible for those with the money and the will. The barriers are political, economic, social.

The powers of technology are themselves neutral. They contain the seeds of happiness and health, or chaos and destruction. Man must choose how he uses them. The problems were spelled out by U Thant, the late UN Secretary General, in 1969: the world has perhaps ten years left, he said, in which to subordinate ancient quarrels between nations and launch a global partnership to curb the arms race, improve the human environment, defuse the population explosion, and supply the required momentum to world development efforts. The tasks are clearly defined. We must now set to work to tackle them.

2 · The Electronic Miracle

In electronics, the near-miraculous child of electrical technology, Man has an excellent means of exploring virtually all his physical environment from the infinitesimal to the infinite. Neither time nor distance offer many barriers to electronics: at one end of the spectrum the electron accelerator permits him to probe the innermost structure of the atom, and at the other the radio telescope enables him to pick up signals from the remotest depths of outer space. Similarly with time. While some electronic circuits are capable of acting in a thousand-millionth of a second, other electronic devices can store information for as long as we need it—for many years, if required. In fact, by providing the means to amplify, select and control electric currents, radio waves and other forms of electromagnetic radiation, electronics has endowed Man with new powers that refine and extend his five senses. Electronic sensors, actuators and circuitry can perform the same kind of actions as human nerves, muscles, brain.

It is this versatility that is the most fascinating and valuable aspect of electronics, and enables it to interact with and improve so many other technologies. To understand why this is so we need to look a little more closely at what it is. For our purposes, it is enough to say that electronics constitutes methods of handling electricity on a small scale outside normal conductors. That is an over-simplification, but it will suffice. As the word implies, it is the practice of handling electrons, electrons being one of the three basic types of elementary particles of which matter is composed. A flow of electrons constitutes an electric current, and an accumulation of electrons an electric potential. Of course, the handling of electrical currents and potentials is also the business of electrical engineering, and since electronics and electrical engineering grew up side by side, it is impossible to draw a hard and fast line between them. Nevertheless, electrical engineering generally deals with electricity in normal conductors (chiefly metals, such as copper and aluminium) and may involve machines weighing tens of tons, and handling currents of thousands of amps. Electronics is usually concerned with small currents—measured perhaps in millionths of an amp—flowing in a vacuum, in gases and in semiconductors

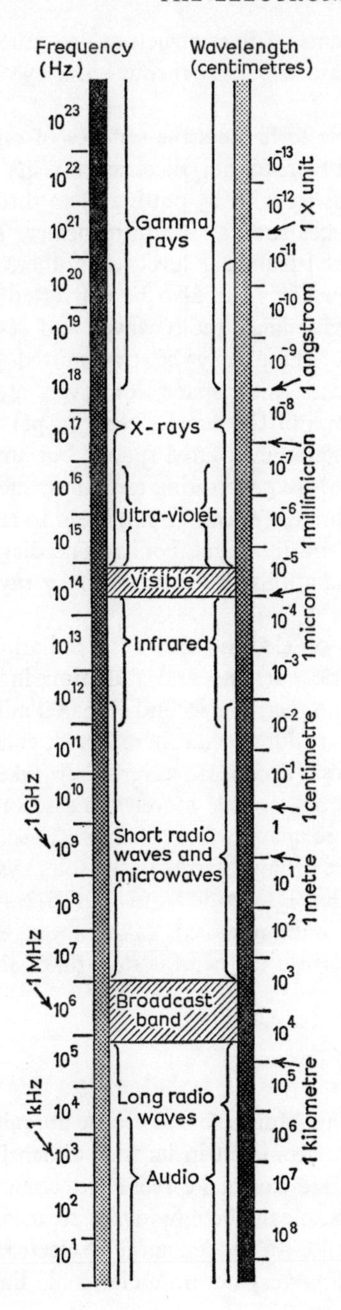

Diagram 2.1
The electromagnetic spectrum

(like silicon), and with components of finer structure—so much so that 700 circuits on a single base will pass through the eye of a needle.

The consequences of being able to harness the services of one of the basic particles of matter are profound. Electric currents and potentials can be varied almost as we wish by putting them through various circuits, or through devices such as the transformer. They can be interconverted, transformed from one level to another, subdivided and augmented. Electron flows can also be converted into different forms of radiation—light, heat, radio waves and X-rays among them—and back again. These may be transmitted over distances measured in light-years into space, or over atomic spacings. Electron flows, and flows of the heavier charged particles known as ions, can be induced across evacuated spaces, but are cut off by air and denser materials. More penetrating radiation, such as X-rays and γ rays, goes deeper, and hence can be employed to reveal underlying structures in animal and inorganic bodies. The diagram 2.1 shows the range of these radiations, extending from γ rays to radio waves.

In earthly terms, the action of electron flows or radiation is practically instantaneous. Electric currents and radiation in free space travel at 30 thousand million metres a second (186,000 miles a second) so that in a second they could, if suitably reflected, encircle the earth rather more than seven times. However, they take an appreciable time to reach the moon—a little more than a second—which is why there is a delay before answers from spacecraft near the moon return to earth. From more distant objects in the solar system the delay is correspondingly greater. Radiation from the sun, travelling 150 million kilometres (93 million miles), takes about eight minutes to reach the earth. That from the nearest stars takes about four years.

The Electronic Yardstick

By dint of his technological cunning, Man is able to use the ubiquitous electron as a yardstick for his environment in its four dimensions, three in space and one in time. He can make these measurements with extreme precision. Often the means he employs is to measure the time taken by an electrical pulse—a sudden surge of electrons—to travel between two points, using very accurate electronic timing circuits. Since the speed of the pulse is a known quantity, the distance it travels can be directly related to the time it takes. This sort of reasoning can be adopted for ascertaining astronomical or microscopic distances.

One of the most valuable things that electronics has been able to do for measurement is to make it much more precise. Again, it is a consequence of using the smallest of the particles of matter as a yardstick. For instance, take the definition of frequency standards used as the basis for international comparisons of time (frequency is just the reciprocal of time). For many years, Man based his time standards on the time that the earth takes to rotate—the mean solar day—defining a second as 1/86,400 part of the mean solar day, and using clocks to measure the day. Then he found a more accurate standard 'clock', in the form of quartz crystals. He discovered that a quartz crystal oscillates at a definite, very accurate frequency which is set by the crystal's dimensions. This allowed him to measure frequency to one part in 100 million. Later, he found an even more precise measure, the caesium beam 'atomic clock'. This emits light of a definite frequency that is used as a standard against which to compare others, permitting him to take his measurements to one part in 10,000 million. Further time and frequency standards based on other atoms—like hydrogen or thallium—may be used in future. Now that technology has attained such accuracy, it is possible to measure time intervals as short as a nanosecond (one thousand-millionth part of a second), and, exceptionally, a picosecond (one million-millionth part of a second). These minute intervals may be used as almost routine measures of machine performance: a computer's speed, for example, may be judged by how many nanoseconds it takes to carry out a given operation.

Electronics has in addition made Man's perception of distance much sharper. The ordinary microscope enables him to discern objects about two hundred-thousandths of a centimetre across. The electron microscope, which 'shines' electron beams at miniscule objects in the same way as an ordinary microscope shines light beams, gives him much better acuity, and refines that a thousandfold, to just over two hundred-millionths of a centimetre. Besides this, electron beams can penetrate matter more strongly than light can, passing through thin metal foils, for instance. They can thus be used to delineate the internal structure of thin specimens of metals, and animal and plant tissues. Recent advances in technology are increasing the electron microscope's power: the latest type, such as that shown in Plate 2.1, has an accelerating voltage of 1 million volts to impel the electrons towards their target. It enables scientists to scrutinize thicker specimens of metals, powders, etc. than was possible with standard instruments of lower voltage, and also to watch whole living cells at very high magnification. In conjunction with photographic magnification, the electron microscope permits us to 'see' objects a million times larger than in real life, and make out the shape of bacteria and large molecules with surprising clarity. In a modification of this instrument, the scanning electron microscope, the electron beam scans the surface of the specimen, producing a striking three-dimensional picture of microscopic structures (Plate 2.2). Another variation, the X-ray microanalyser, allows specimens to be analysed by identifying the X-rays that are emitted when the electron beam strikes the specimen. An additional instrument, the field-ion microscope, enables us to pinpoint individual atoms on the end of a very sharp metal needle, and photograph the pattern made by single metal atoms, which show up as points of light. The structure is magnified typically five million times.

Although Man cannot see further than this, he can probe deeper into the structure of matter within the atom by shooting nuclear particles at stationary targets and analysing the particles that are ejected. In the particle accelerators that are used to accomplish this shooting, protons, electrons and other particles are accelerated under high voltage and in a very high vacuum—so that they are not stopped by colliding with air molecules—and hurled at the target. A common way of observing the emerging spray of particles from the target is to watch their tracks in bubble chambers of liquid hydrogen, photographing the tracks for subsequent analysis. The ultimate aim is to discover the symmetry that rules the whole disordered host of particles populating the realm of nuclear physics, and to explain the nature of the interactions between them. When he understands them, Man can utilize them, as he has done with the electron, the proton, and radio waves and other radiations.

At the other end of the scale, radio-telescopes peer into the voids of space, searching out the remotest galaxies, recording their presence, their movement, and their strange emissions of radio waves and other radiations, whose occurrence poses new challenges to present theories of the universe and of creation. The technology of electronics thus assists Man as he gropes towards an understanding of two of his most basic and most intractable problems: the nature of matter, and the mechanism of creation.

Electronics has far more than an intellectual attraction, however. Because of its great versatility, it has invaded many other branches of science and technology in the past decade and is now firmly implanted at the heart of modern measurement, communication and control. Without it automation would be impracticable. It provides the 'building blocks' from which many of our commonest devices—transistor radio, television set, gramophone, tape recorder, telephone and the rest—are built. Not only the common, everyday things but also such items as computers, industrial instruments and control gear, electron microscopes, heart pacemakers, and rocket guidance systems owe their existence to electronics. In the last twenty-five years the business community and government have recognized this value, and now electronics is one of the fastest-growing parts of industry, expanding at around 10 per cent a year. It is providing many new jobs and new opportunities, not merely for those working in the factories making electronic components, but also far more in the plants turning out radios, television sets, tape recorders, telephones, and a host of products that have been made cheaper or more readily available because of the influence of electronics technology upon them. Besides these people, there are others who are designing circuitry and devices, carrying out research and development, teaching and training others in the skills of electronics, and so on.

Towards the Microscopic World

Paradoxically, electronics is continually getting more out of less: the actual devices are being made smaller and smaller, and while their costs fall and reliability improves, their applications multiply. This fact has big implications for the industries that make them and those that use them, and eventually and most of all for the people who use them—since it puts within the reach of the ordinary man cheap, reliable devices which were previously purchased only by the wealthy few.

Miniaturization of the circuits has been the most important factor. The reduction in the size of electronic circuits has been steady and impressive. Not long ago one could regard a packet of cigarettes as a reasonable comparison for the size of electronic components.

Then the components were miniaturized and small coins made a more apt comparison. Now the components are smaller still, and dwarfed by grains of salt. They have shrunk by a factor of ten roughly every five years.

The three major steps in miniaturization correspond to three stages in electronics technology (Plate 2.3). At first, valves and other components were made individually, then assembled and wired up one at a time on a frame, using a soldering iron and other hand tools. Next came printed-circuit boards and separate solid-state devices, notably the transistor. The transistor, discovered by research workers at Bell Telephone Laboratories in the United States in 1947–48, performs within the space of a few square millimetres the same tasks as a valve many times larger, but at much lower voltage—typically six volts instead of 200 volts. It consumes less power, is less complicated and has less in it to go wrong.

By 1953 transistors could be produced cheaply, in large quantities. Other components were miniaturized too, and although they still had to be assembled one by one on the printed circuit boards, they could be connected simultaneously by dipping the complete board into a bath of molten solder, obviating the laborious and finicky work with the soldering iron.

The perfection of the transistor opened the door to the third stage: the production of integrated circuits, not fabricated from separate components that were afterwards connected on some sort of frame, but created within the semiconductor base itself. The general definition of such a circuit is one that is made as a single entity and that cannot be taken apart and reassembled. The various components—transistor, resistor, etc.—are fabricated within different regions of the semiconductor by chemical and physical processing. The integrated circuit represents the present stage, and, so far as can be seen, the final one. Apart from its obvious size and weight savings, it is more reliable, faster, less prone to damage and potentially cheaper than its counterpart made up from separate components. The methods of making such circuits permit fantastic numbers of components to be packed into minute areas: five years ago an integrated circuit incorporating the equivalent of ten conventional components was doing well; now circuits with the equivalent of 200 to 500 components are being made. Laboratory work has already produced devices with about 250,000 components to the square inch. So rapidly do designs and ideas change that even

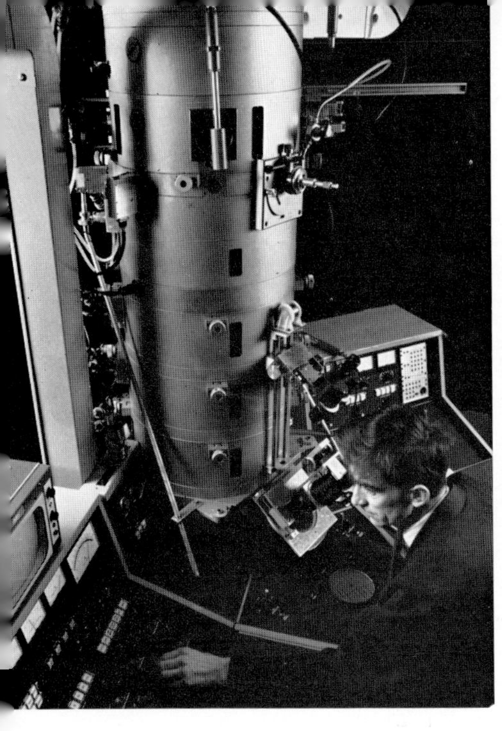

2.1 (*left*) Million-volt electron microscope, now installed at the U.K. Atomic Energy Authority Establishment, Harwell

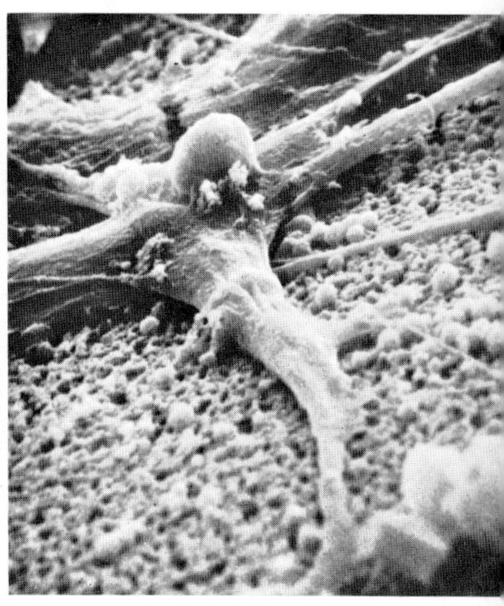

2.2. (*right*) Three-dimensional picture of a malignant cancer cell, taken by scanning electron microscope and (*below*) crystals of an alloy similarly viewed

(*a*) chassis with valves

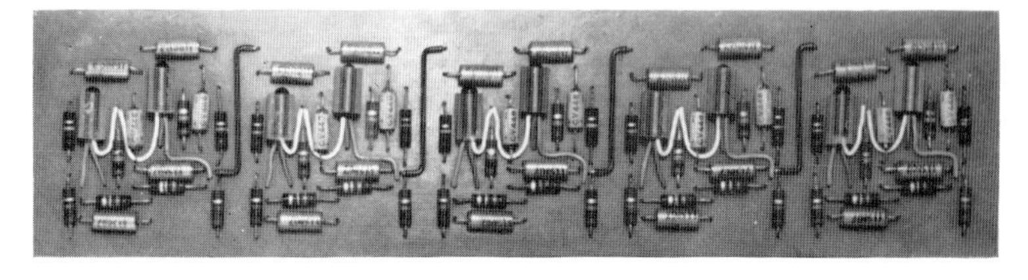

(*b*) printed circuit board

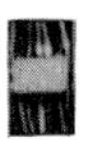

(*c*) semiconductor integrated circuit

2.3 Miniaturization of electronic circuitry: how a particular circuit has shrunk over the years

2.4 Integrated circuit assembly line

2.5 Experimental solid-state colour television set

3.1 Basic oxygen furnace in a northern Japanese steelworks

3.2 Part of the control room of a modern chemical plant

4.1 A typical large computer installation

4.2 Video computer terminal displaying computer-generated information

4.3 Designing integrated circuits on a video terminal with the aid of a light pen

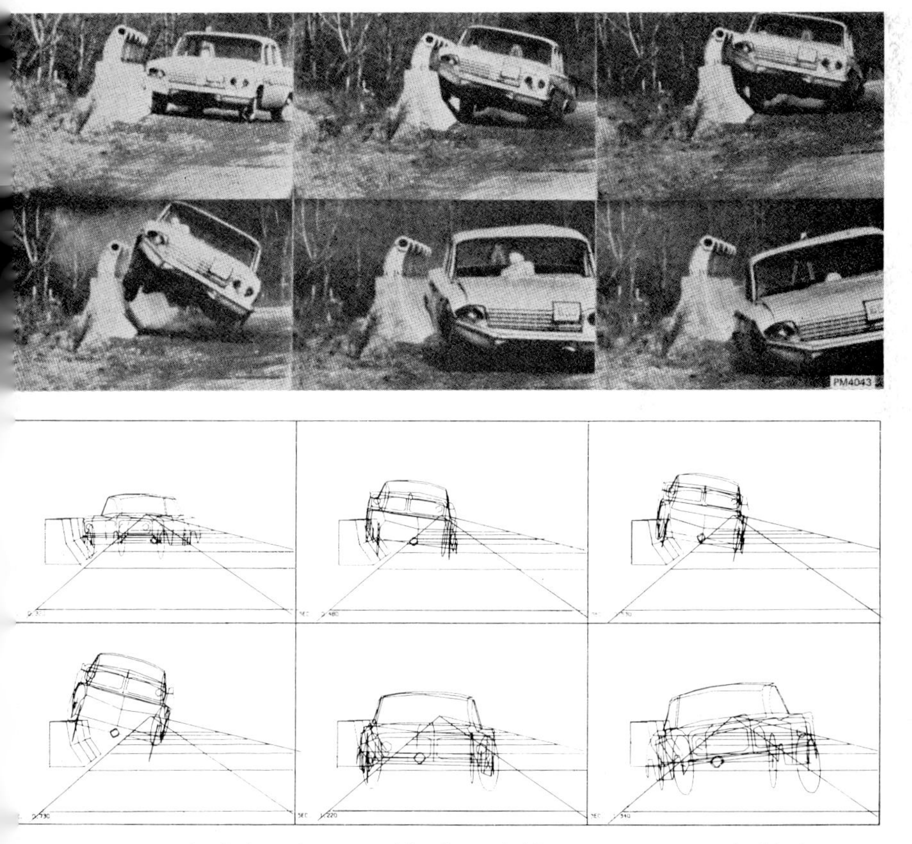

4.4 Computer simulation of a car crashing into a bridge parapet, compared with the actual event

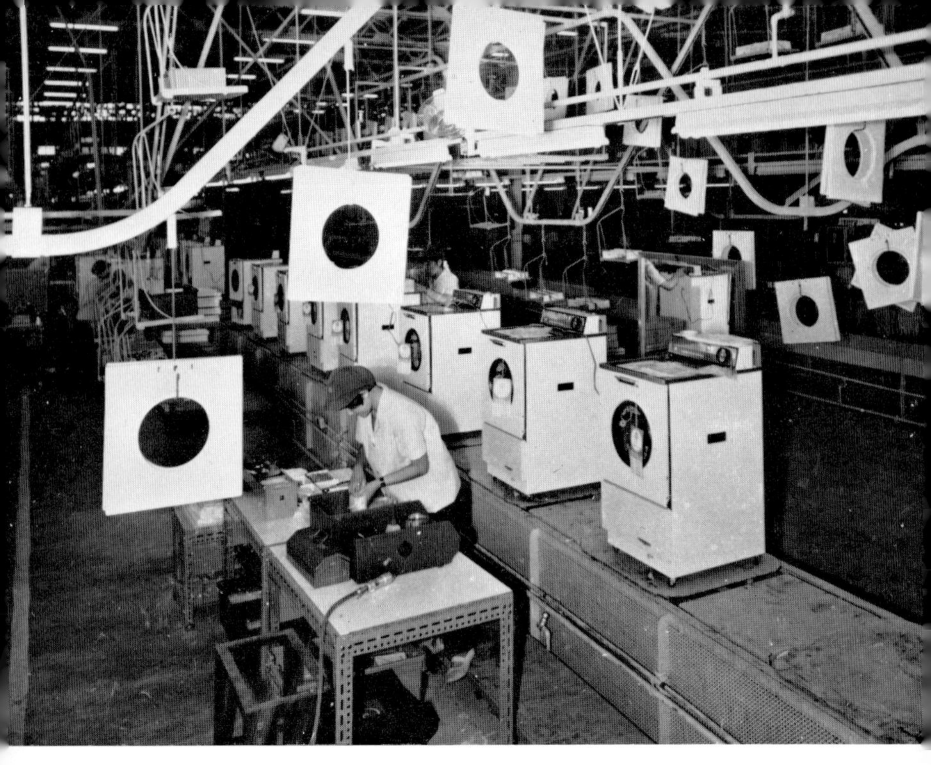

5.1 Highly automated Japanese factory, designed to produce 15,000 automatic dishwashers a month

5.2 Assembly line at a Japanese factory whose chief products are radiograms, record players, hi-fi components, and speakers

5.3 Ultrasonic weld examination of pipeline joints

5.4 Modern mass-production of cars on the assembly line

6.1 Installing an inland pipeline to bring natural gas from the North Sea reserves into Britain

6.2 One of the latest of Britain's nuclear power stations—at Oldbury-on-Severn

6.3 Heavily shielded 'cells' in which radioactive materials are handled by remote-control manipulators

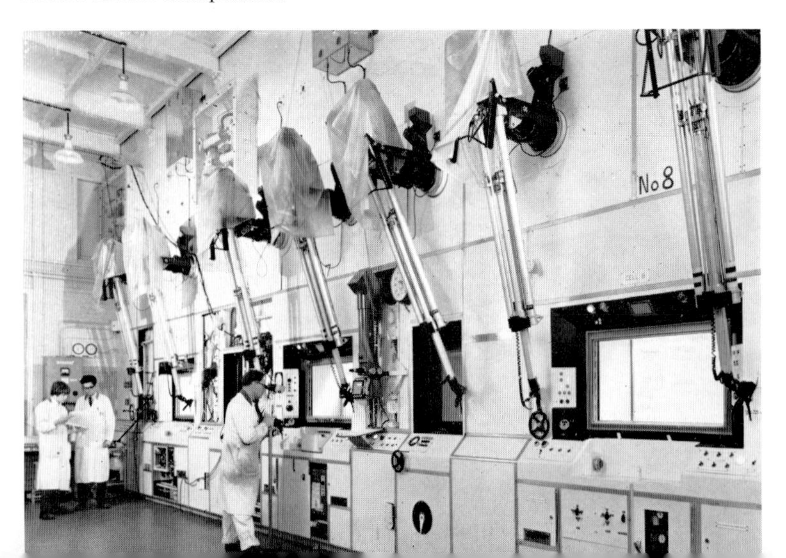

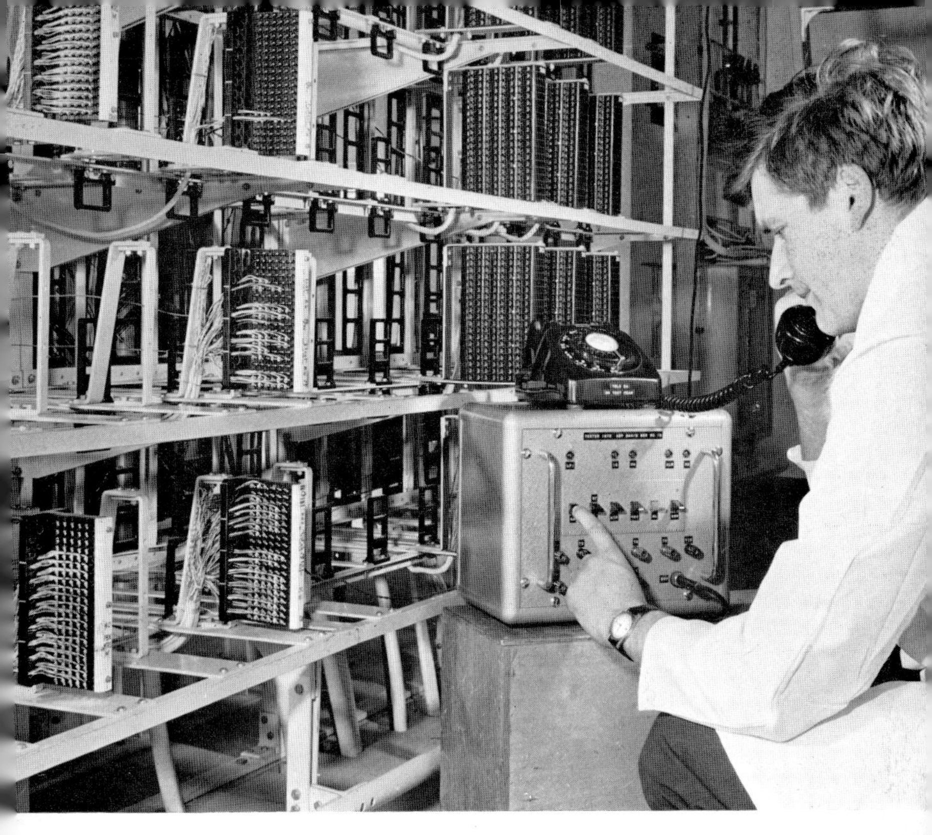

7.1 Checking the operation of an electronic exchange

8.1 The Galaxy, largest aircraft in the world in 1971

8.2 Cutaway model
a Mini, the city car
the 1960s

8.3 Closed-circuit television gives a controller a continuous picture of traffic conditions—here, traffic crossing under the River Thames through the Dartford Tunnel

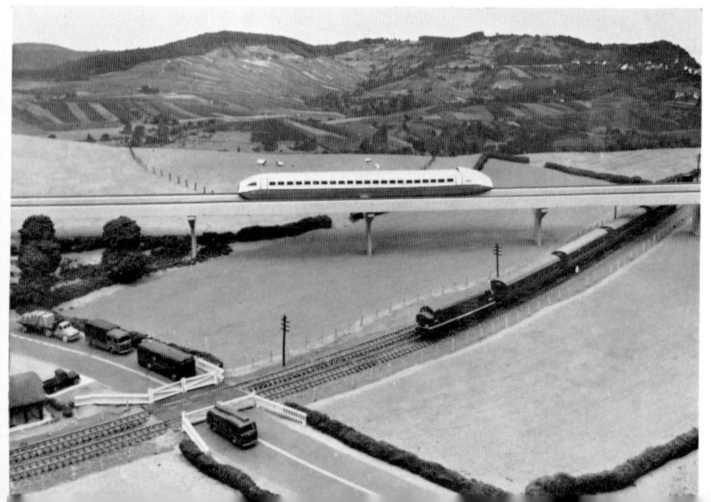

8.4 Model of a typi
tracked hovercraft
installation

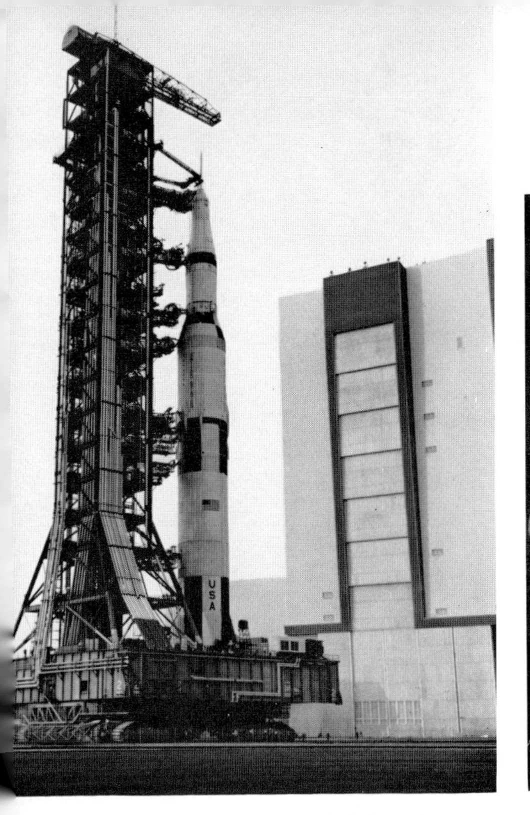

The U.S. Apollo 14 launch vehicle being taken
m the colossal assembly building to the launch

9.2 Early Bird, the world's first commercial communications satellite, gets final adjustments in preparation for vacuum tests

9.3 Astronaut David R. Scott during a space 'walk' in the Apollo 9 mission

9.4 Huge vacuum chamber simulates conditions in outer space or on the moon for testing spacecraf

10.1 Highly manoeuvrable swimn sled for shallower waters

10.2 The research submarine *Deep Quest* for deeper exploratio

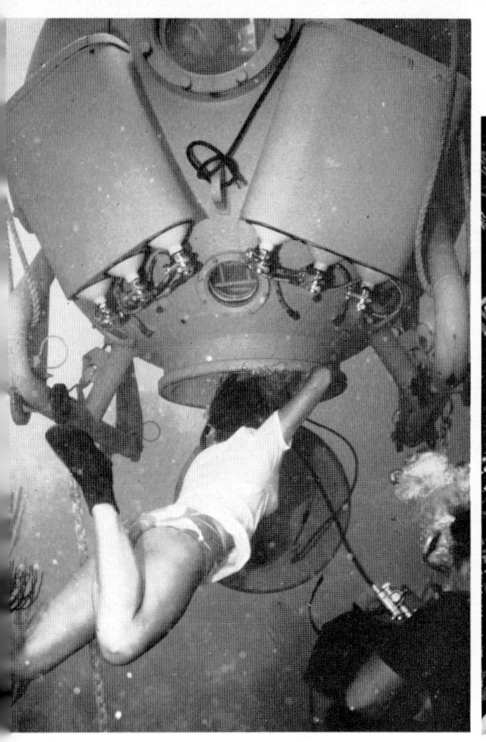

10.3 Female aquanauts, living beneath the Caribbean for two weeks, leave their underwater home to begin a scientific experiment

10.4 Test subject wearing a pressurized space suit under simulated weightless conditions in crystal-clear water tank

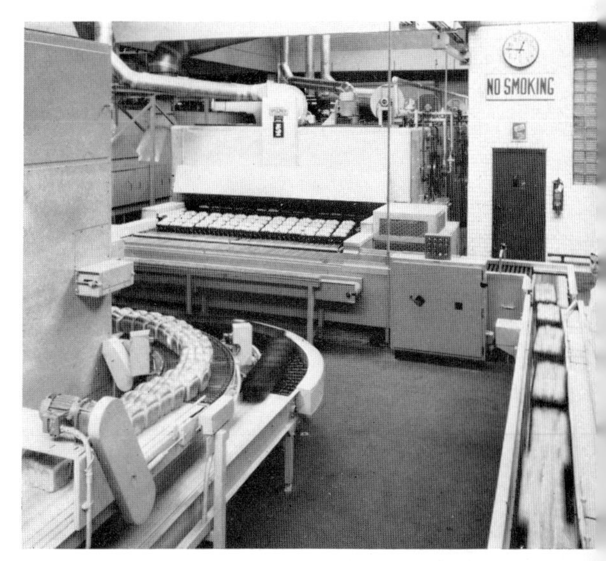

11.1 A highly mechanized bakery

12.1 and 12.2 Assembling prefabricated units on a building site, using the Bison wall frame system

12.3 The new town of Cumbernauld, Scotland

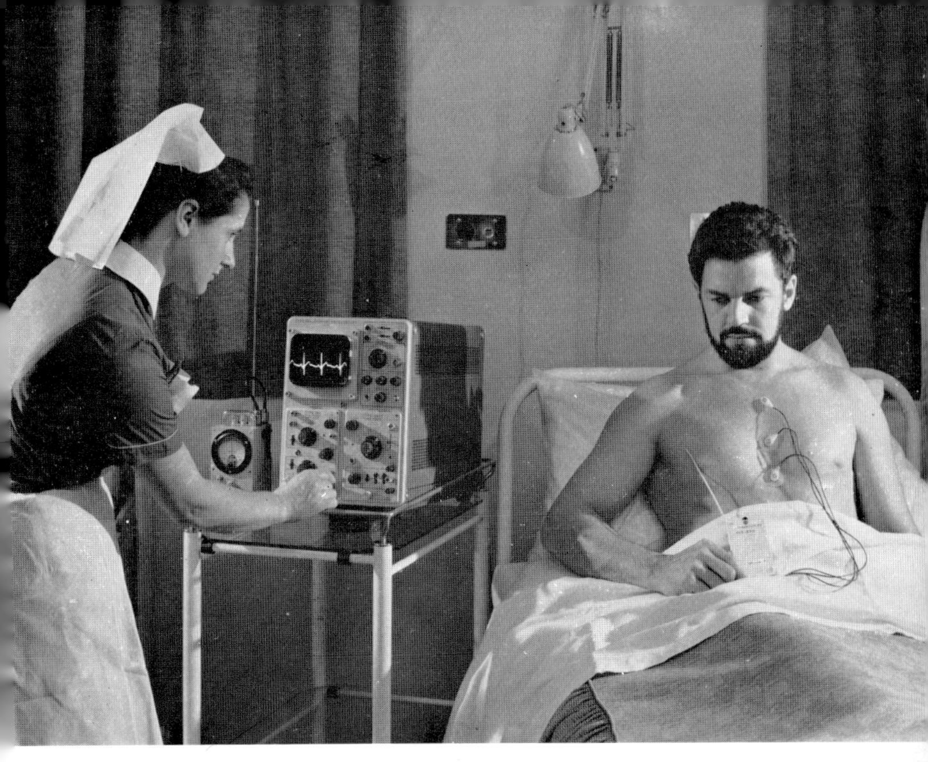

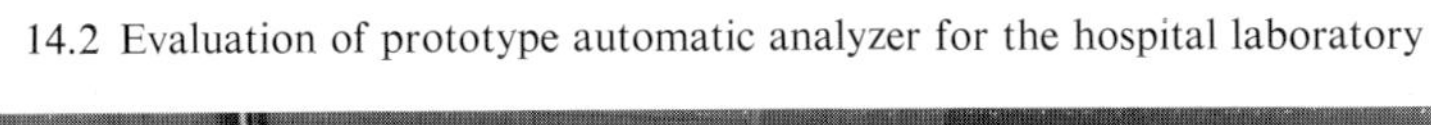

14.1 Monitoring a patient through electrodes which pick up bodily electrical activity

14.2 Evaluation of prototype automatic analyzer for the hospital laboratory

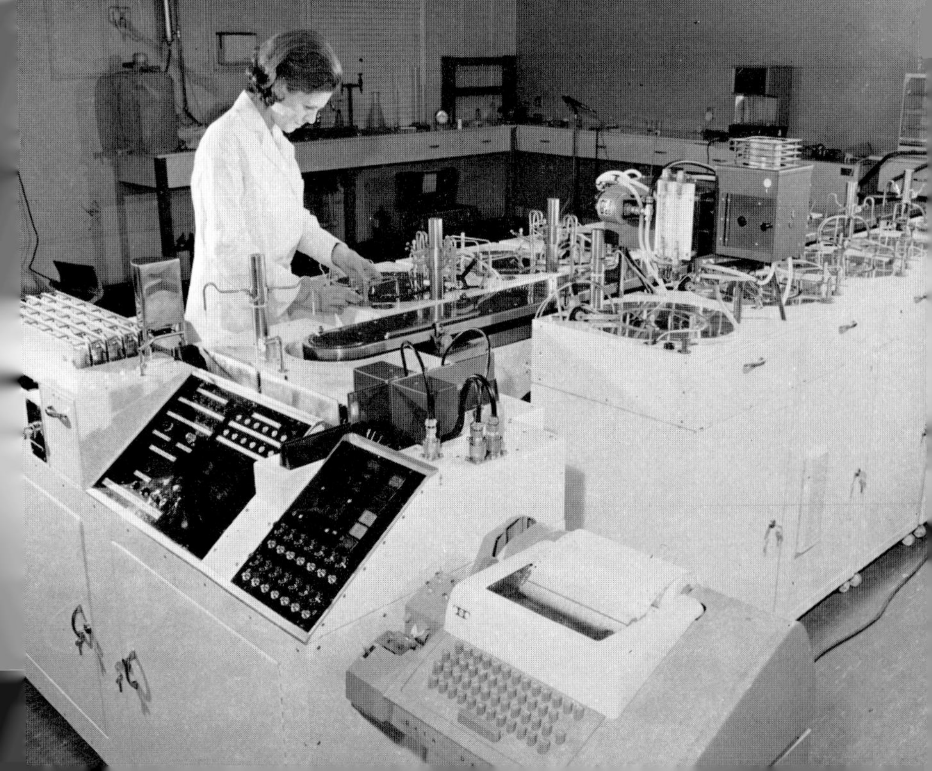

14.3 Modern pharmaceuticals manufacture

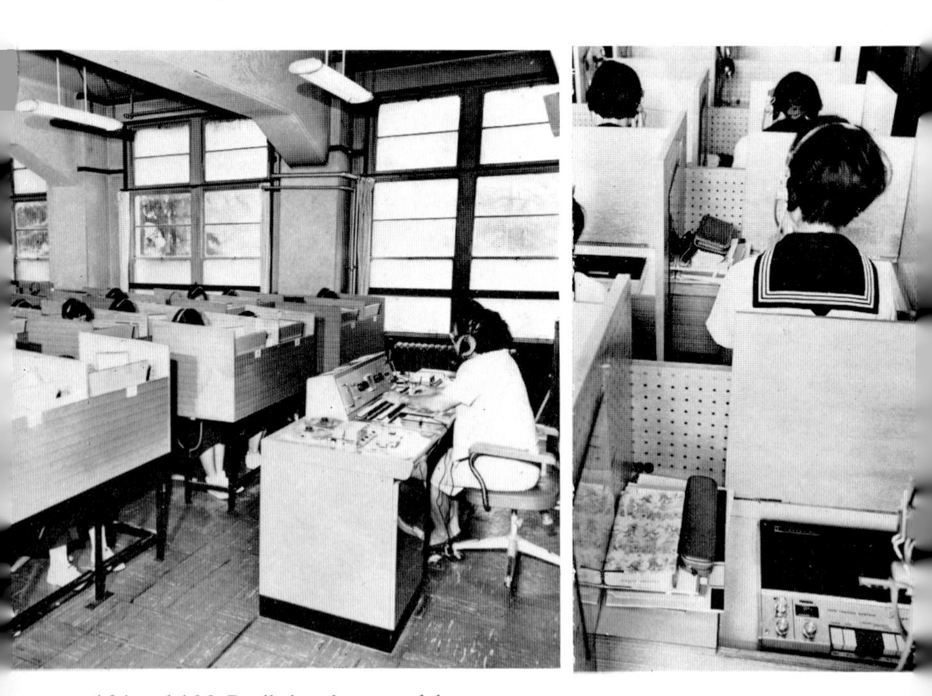

16.1 and 16.2 Pupils in a language laboratory

experts decline to predict what integrated circuit technology will look like in five years' time.

One of the great benefits of the miniaturization process is that it makes electronic devices more reliable—a very important factor as the devices become more involved. The portable radio of the mid 1950s, for instance, had no more than about 100 components. Electronic devices in the early 1960s such as computers and satellites could run up to between 10,000 and 100,000 components. The most complex computers of the early 1970s may have of the order of a million components. At each step, the chances of breakdown multiply. The more components there are, the greater the chance is that one will put the whole machine out of action. The cost of maintaining such complex machines can be substantial, perhaps ten to 100 times the original cost of the machine during its useful life. Fortunately, miniaturization has improved reliability enormously, mainly through eliminating many massive components and a great deal of wiring. As a result, the failure rate, in terms of failures of components per hour, has decreased by a factor of 1,000 in the last ten years.

The new, reliable circuits are finding applications in many areas that were quite unsuitable for the earlier circuitry. The contemporary generations of computers are based on integrated circuits. One of the very promising areas in which integrated circuits may be applied is motor vehicles, especially cars, whose electrical systems have altered little since the mid 1930s, apart from the introduction of alternators with solid-state rectifiers to replace the dynamo. Many cars, however, are fitted with transistor radios, providing a precedent for further solid-state devices. Plastic-encapsulated, solid-state devices could replace conventional components in the ignition system, instruments, windscreen wiper, and so on. In the home, integrated circuits could be widely used to control such things as washing machines, dish washers, refrigerators, electric cookers, colour television sets, record players and radios. In telecommunications, medical electronics, data keyboards for computer terminals, and scientific instruments and navigation aids, the tiny circuits will find ready markets. In many instruments, the use of integrated circuits offers the first reliable and economic way of presenting the readings in an unequivocal digital form (a series of numbers) instead of as a needle moving round a dial, which is a frequent source of error. Then there is espionage: tiny microphones, capable of overhearing conversations across a crowded room, radio sets inside cigarette lighters and transmitters hidden in the 'cherry' of a cocktail owe their clandestine existence to the integrated circuit.

Already integrated circuits are big business. As early as 1967 42 per cent of the television sets and over half the high-fi equipment made in the United States incorporated at least one integrated circuit. All such units were expected to be using the circuits by 1972; and by then the demand from the consumer field in the United States alone was expected to have risen from a level of about two million a year in 1968 to well over 200 million a year. The early research work carried out by Bell and other firms in the 1950s has really paid off. What started as a search for the scientific explanation of the action of the 'cat's whisker' crystal radio has grown into a major industry based on silicon—the cat's whisker crystal was silicon carbide—and the closely related material germanium. This industry is enjoying some of the wildest growth rates ever seen, some firms expanding at rates of more than 30 per cent a year—and some of the worst failures occurring among those companies that do not have the necessary competitive edge.

The manufacture of integrated circuits is a prime example of a high-technology process, demanding specialized machines in which extreme conditions are counted as routine. The necessary chemical and physical treatment in which various sections of the semiconductor—usually silicon—are deliberately 'doped' with small, controlled amounts of various impurities to give them the right electrical properties, is carried out in a number of steps. Insulating material and conducting material are deposited in carefully delineated regions to make up the required circuit pattern. The circuits are usually made on a slice—or 'wafer'—cut from a single crystal rod of very pure silicon. On this slice, typically 5 cm across, many hundreds of circuits are formed, interconnected perhaps by minute streaks of gold or aluminium, laid down on the surface under high vacuum. They are so small that they have to be inspected and tested under a low-power microscope, and handled by dexterous girls with tweezers and good eyesight (Plate 2.4). Complete circuits are put into a holder or case and fitted with wires of normal dimensions to connect them to ordinary components in the outside world.

Manufacturing conditions are stringent. In some processes the slices in quartz trays are heated to a temperature of 1200°C, which has to be controlled to half a degree. In others they are heated under high vacuum. In most parts of the manufacturing plant strict cleanliness is essential to prevent stray dust from contaminating the circuits. Air is filtered and the personnel wear special clothing to stop dust and hairs coming near the precious circuits. Even so, reject rates are high. The circuits are handled at virtually every stage, and at each there is a chance that the operator will break one of the brittle grey slices or, worse, drop a tray of hundreds of them. Those circuits that complete the manufacturing process are tested, still on the slice, by a machine that brings a battery of tiny needle probes to bear on each separately. Rejects are thrown away. In the case of some complicated designs manufacturers are pleased if they can sell 10 per cent of the initial number. The materials themselves are relatively cheap compared with the value that all this processing confers on the finished circuit. Roughly a third of the cost of the finished article is accounted for by the semiconductor crystal and what it contains in the way of 'doping' compounds, a third by its container, and a third by the cost of testing it.

Another stage on from separate integrated circuits is large-scale integration, which involves arrays of circuits on one slice of silicon a few centimetres across. The circuits are permanently linked together by interconnections corresponding to conventional wiring that are laid down on the silicon itself. The interconnected circuits are sold in a single package. Normally each one would be separated and tested, faulty ones being thrown away, and good ones encapsulated in plastic, tested again and sold singly. It is logical to leave all the circuits together on the slice and connect them in different ways according to the kind of function that the total circuit is supposed to perform.

One way of doing this is to test them automatically and then have a computer generate the required connection layout on a cathode-ray tube so as to avoid the faulty ones and connect the good ones. The pattern on the tube is photographed, transferred to a photographic mask, and then employed to make the required metallic connections on the silicon slice.

Large-scale integration (LSI) will mean further price reductions. The manufacturing cost of integrated circuits depends chiefly on the number of steps involved, not on the number or disposition of components in a given area. It should cost no more—or little more—to make 10,000 transistors on a stamp-sized silicon chip than to make 1000 or 100. Theoretically, then, the user could buy an LSI package that would do the job of ten ordinary integrated-circuit packages, but costing only as much as one of them. As costs fall, it gives the manufacturers who use them in their products the chance of introducing more integrated circuits and thus greater refinements in their products for the same price. For example, the manufacturer of colour television sets could introduce an automatic frequency-control circuit: a colour set reproduces colour best when it is tuned precisely to the centre of the television-frequency channel, and the tuning is delicate and difficult to do manually. The alternative course for manufacturers is to keep the product in the same form but cut its price, or rather hold the price steady amid general inflation.

Large-scale integration therefore opens new vistas in electronics, in which circuits of great complexity, capable of performing such operations as adding, dividing, and remembering, can be constructed on tiny silicon chips, each no larger than the word 'the' in this print. A computer logic unit can fit on a silicon chip only about one third of a centimetre square. With its twin effects of raising reliability and lowering costs, the technology of miniature electronics devices will permit many more people to own radio and television sets, telephones, and other items, and so permit them to be better equipped to communicate with the outside world. Looking beyond current uses, the day of the wristwatch television, electronically guided cars, computer terminals in the home, and even robot toys, is not far off.

However, other electronic components are more difficult to miniaturize. At the moment it is hardly worth using integrated circuits for the normal television set when the size and shape of the set is determined by the bulky cathode-ray tube (see Plate 2.5) and its power supplies. Similarly, the size of the radio or high-fi set is dictated largely by the size of the loudspeaker. If electrical engineers are determined to solve the problems, it has taxed their ingenuity sorely. The cathode-ray tube, for instance, is a basic device of electronics that has changed little since the days when J. J. Thomson, at the Cavendish Laboratory in Cambridge, employed a rudimentary tube in his discovery of the electron and Karl Ferdinand Braun of the University of Strasbourg made the first cathode-ray oscilloscope to aid his study of alternating voltages. That was in 1897. Since then the cathode-ray tube has done yeoman service, and today it forms the tube of the television set, the radar set and the oscilloscope, providing the means of converting electron flows into visible signals. Though much refined, it is still essentially an evacuated glass envelope across which electrons are shot under high voltage, to appear as a glowing trace on the tube's phosphor coating. Electrodes that are needed to deflect the electrons' flight in order to create a picture are sealed into the glass or wound round the outside. The tube is a delicate, complex, expensive piece of apparatus, quite apart from its bulk. Much work is directed at reducing its size, delicacy, complexity and cost. One technique is to 'squeeze' the tube into a smaller space—creating a so-called flat tube. Other investigators adopt a more revolutionary approach and generate the display with an array of individual light sources equivalent to tiny electric bulbs, which are switched on and off in patterns to build up the picture. This is how one kind of wristwatch television would work. Another approach—and so far the nearest to commercial application—is to have a flat sandwich of phosphors or liquid crystals activated by two intersecting grids of wires. The picture is created by sending electrical pulses down both sets of wires: where two wires intersect the pulses generate a spot of light. Eventually the cathode-ray tube is likely to be replaced by stronger, simpler devices. But there is a lot of life left in it yet.

Electronics and Light

Electronics is a young technology. To see one that is even younger we need do no more than turn to the related field in which electronics interacts with light, or opto-electronics, as it is called. The laser, one of the curiosities of the modern scientific world, stands supreme in this field. In its early days it was dubbed 'a solution in search of a problem', but technology swiftly takes over from science these days, and only some ten years after the effect was demonstrated in May 1960 the laser had found several hundred uses in laboratories, hospitals and factories. Many more uses are being found.

What the laser does, in effect, is generate an extremely pure, intense beam of light (or other such types of radiation, as infra-red or ultra-violet) by an energy amplification process within a solid, a gas or occasionally a liquid. Its counterpart, the maser, carries out a related amplification process to produce microwave radiation of the kind needed for radar and various modes of communication. The laser's beam can be focused down to a narrow, almost parallel beam of light. So narrow is the beam that when projected on to the moon, 386,000 km (240,000 miles) away, it spreads out to illuminate an area only 3·2 km (2 miles) across. The laser has other showy tricks such as piercing through metals or even diamond amid a shower of sparks. Some people still think of the laser as a potential 'death ray', and indeed the United States for one now has experimental laser weapons in operation.

In the civil field, the laser brings great accuracy to such tasks as surveying, measuring to a matter of centimetres in a distance of several kilometres. It has been used in this way to measure the distance between the earth and the moon more accurately, after American astronauts set up laser reflector panels there in 1969. In other varied tasks the laser can indicate very accurately the position of the work-table on a machine tool; or the trembling of a building during an earthquake; or the speed—which is change in distance over a measured time interval—of a train or aeroplane; or determine the size of particles in the air, a matter of great importance in the ultra-clean areas in integrated-circuit plants mentioned earlier.

Lasers generating infra-red radiation can illuminate the scene at night with these 'invisible' rays strongly enough for a sensitive detector to produce a picture on a television screen in complete darkness. On the screen the picture appears almost as though it were daylight. On the battlefield a sniper equipped with a laser mounted on the rifle can virtually see in the dark: so can police or security forces watching for car thieves or intruders. One of the most important applications—potentially—is in communications, since messages can be sent along its beam rather as they are sent along a cable. The laser beam, however, can carry many more messages.

Unfortunately the beam is scattered by rain, fog, mist and even the air, so over long distances the beam would have to be protected—long evacuated pipes have been suggested—or carried in some sort of light guide, possibly fibre optics guides. These are basically bundles of very fine glass fibres, only 0·1 mm or less in diameter, about as thick as a hair, similar to those made into industrial cloth, or plastic reinforcement. They are of a peculiar type, though. Each fibre has a core of one type of glass surrounded by a sheath of another type, so that light shone in at one end is reflected along inside the fibre. A bundle of them is extremely flexible and can transmit light—and hence an image—even when the bundle is tied in a knot. This makes it possible to 'pipe' light round corners, and into all sorts of novel situations—inside the stomach, for example, to inspect damage wrought by disease. Theoretically, certain of these fibres would make good carriers of telephone and television signals. So too would microwaves.

The discovery of the laser has roused fresh interest in optics, and especially in non-linear optics, where light does some very odd things. Take holography as an example. The effect was predicted by Professor D. Gabor of Imperial College, London, in 1947, but its experimental verification had to wait until the laser came thirteen years later. A hologram is a kind of three-dimensional photograph of an object taken in laser light. It seems to be a meaningless jumble of lines and blobs on a photographic plate when viewed in ordinary light; when illuminated in the pure light of a laser it springs into life, reproducing a three-dimensional view of the object. As the viewer moves his position, it is as if he were looking round to the side of the object: if there are several objects on the hologram, they appear to change their relative positions.

Holograms can be stored on film, or in special crystals, whose optical properties are changed during the 'photographing' of an object in laser light. In fact, up to 1,000 holograms can be stored at once in a single small crystal by rotating it a fraction of a degree between one 'picture' and the next. The different holograms can then be reclaimed in three dimensions. This of course has immense implications for systems of information storage and retrieval. It might make an optical filing system even more compact than microfilm, and it raises thoughts of holographic cinema, and decorative panoramas on the living room wall that could be changed almost endlessly at the flick of a switch.

Holography can also pick out complex movements in car tyres, machines, bridges, and other moving or stationary structures. Movements usually invisible to the naked eye show up as dark and light areas in the hologram, and can be preserved in this form for analysis. Holograms can be made with other forms of radiation, too —that is, not with a laser at all. Those made with sound may turn out useful in oceanographic work, as sound waves penetrate much further than light in water, and in medical diagnosis, because they are less harmful to the body than X-rays but penetrate the tissues in a similar way. Others may be fashioned with the aid of microwaves, and perhaps even X-rays, in order to obtain high-resolution pictures of microscopic objects, although there are considerable experimental problems involved. The field is widened considerably by the fact that holograms can be created within the computer, being printed out as a digital pattern which yields a three-dimensional image when illuminated by laser light. This technique—digital holography—seems an obvious one for displaying designs, for cars say, on account of the great savings in modelling time and cost. The designer could feed into a computer the desirable parameters, such as engine power, weight, capacity, overall weight, shape, handling characteristics and so on, and then produce a digital hologram of the car and view it in three dimensions without having to create anything real at all. It would all exist in the computer's electronic circuits and on film.

The field in which optics and electronics are married is becoming more intriguing as science probes deeper. There are undoubtedly many more offspring of the union to come. Other modes of interaction between electricity and light can produce useful and sometimes unexpected effects. In the conventional light bulb, for example, the filament—generally a doubly-coiled helix of tungsten wire—is heated white hot to cause it to emit light. The heat is obtained as a wasted by-product. There is therefore much interest in obtaining 'cold light' (luminescence) without having to heat the material to such temperatures. In the fluorescent tube light is obtained by exciting mercury gas to give off ultra-violet radiation which causes a phosphor coating on the tube to glow. Attempts to get cold light by passing an electric current through solids have met with less success. However, rapid advances in semiconductor technology—particularly experience gained in the mass production of integrated circuits described earlier—has recently brought many more possibilities to notice. A voltage applied to certain carefully prepared semiconductors, notably gallium arsenide, makes them glow. Electricity is converted into light directly and with great efficiency. These semiconductor lamps can be made by the same large-scale integration techniques that were described earlier, and may be made to the same microscopic scale as integrated circuits. This means that arrays of them may be set out in a very small area of semiconductor in the same way that integrated circuits are.

Research on other materials such as gallium phosphide, gallium arsenide phosphide and silicon carbide (which is used as an abrasive in another form) is continuing. All these have been developed as tiny indicator lamps, covering most of the colours in the visible spectrum. Their earliest uses are as indicator lamps for calculators, telephones and electrical instruments in aircraft and elsewhere. One type, for example, displays in sequence the numbers zero to nine in bright red in an area about the size of a drawing pin.

The effect works the other way round, so a semiconductor device can convert light into electricity. The photocell achieves this in many different kinds of situation and forms the basis of the exposure meter, numerous scientific instruments, industrial counting and detection mechanisms and so forth. Large arrays of sizable photocells, deployed around space satellites, draw power from sunlight. Small, compact arrays, which like the semiconductor lamps may be made by the same kind of processes as are used for integrated circuits, are potentially ideal replacements for the human eye in reading and abstracting information from documents at high speed. The technique of optical character recognition is one that human beings are expert at, and it will take some time before machines can match the human performance. Once they can recognize printed—or written—words and figures, though, they will be able to take on a vast amount of dull, routine work at present done by large numbers of office staff.

Electronics can do a great deal in partnership with light, then, and nearly as much with infra-red (heat) radiation, which lies not far away in the electromagnetic spectrum. Until recently the development of infra-red technology was largely shrouded in military secrecy but the wraps are coming off, and civil applications are multiplying. Infra-red systems sense and map the patterns of heat waves that are emitted by all objects but which are invisible to the unaided eye, ordinary film and television camera. In one type of system, a scanning detector converts the heat signal emitted from an object into an electronic signal, producing a television-like image on a display screen. Industrially, such a system is used for detecting incipient 'hot spots'—or points of failure—in current-carrying devices before they actually fail, since the spots stand out against the background in the 'heat picture' obtained of the equipment. It is also widely used in North America and Europe as a medical aid, for diagnosing breast cancers, locating tumours, varicose veins, etc., since it detects the small differences in body temperature around these abnormal areas and records them on film for the doctor to inspect.

Experimental work further indicates that acousto-electronics—the union between sound and electronics—may be as fruitful as that between electronics and light.

Electronics extends Man's powers over the environment down to the smallest particles of matter, and allows him to plumb the remotest depths of the universe for clues to its creation. These powers have not been his for much more than three generations, however. The three pillars on which the edifice of modern electronics is reared are the *cathode-ray tube*, which, as was mentioned earlier, dates from the end of the nineteenth century, the *thermionic valve*, created on the borderline between the nineteenth and twentieth centuries, and the *solid-state devices*, heralded by the transistor, which are definitely part of the twentieth.

These devices and many more form a vital part of modern technology. They utilize some of the most pervasive and fundamental phenomena of the environment, rendering the effects of the natural world in terms that the human being can appreciate. Research in this field and related ones engages some of the world's finest brains, and a sizable proportion of its research effort: it is continually unearthing new effects that widen the scope of electronics still further. Opto-electronics is one. Man is in fact far more familiar with light than electronics; now, with the aid of the laser and photo-electric and other devices, he is putting light and electricity into the same harness to work for him. Such mixed technology has great potential and may well overshadow the straightforward technologies like electronics in the future.

3 · Materials of Progress

Materials Ancient and Modern

In his natural environment Man found bone, clay, wood and stone for making things, and animal skins for clothing himself. But these did not satisfy his creative impulse for long. Over many centuries he painstakingly learned to extract metals such as iron and bronze for structural materials, and discovered how to make cloth from wool, silk, cotton and flax. Iron, steel and some non-ferrous metals such as gold and copper, together with concrete, timber and paper, were made and used 4,000 years ago and earlier—and in some cases very much earlier. Ceramics date back 10,000 years; glass 6,000 years. Some of these are still technology's most prized materials. The only difference is that now they are made and used on a massive scale, with much higher and consistent purity, and in many formulations.

But modern technology involves far more than these. No longer satisfied with the restricted range that Nature provides, Man has created a host of new materials, using technology to bring them to birth. Not all are compounded from the 94 natural elements of which the world is composed: others have been created by Man in the massive radiation flux of the nuclear reactor or the electron accelerator. Materials research is now a major field of study and discovery. By examining materials under the microscope and other instruments (see page 45) we have gained an understanding of the way they are constructed, how faults occur and move through them under load, and how the material realigns itself as it is stressed. We may even observe the images of individual atoms, their position, and how they fit into the bulk structure.

The techniques of X-ray diffraction and electron diffraction (see page 44) enable us to analyse structure in considerable detail, to correlate its measured properties with the structure, and then refine theories so as to make prediction of the properties more reliable.

As a result we are richer in materials that we can choose for technology, and we can use them with more skill and daring. With the aid of research we are much better informed about their properties so that we can predict accurately how they will behave in service and can design structures with more confidence. Each advance in materials helps technology forward another step. Aeroplanes, nuclear reactors, spacecraft and integrated circuits could not have been built without new steels, alloys, ceramics, composites, semiconductors and other advanced materials. Plastics have created whole new families of materials, rivalling both steels in their lightness and resistance to corrosion, and natural materials such as cotton and wood in other fields.

The designer has a choice embracing thousands of materials for his new machines and structures. The engineer nowadays is likely to find two, or even three or four, radically different kinds of materials ready for a given use. In such industries as electronics and aerospace, where cost has not been paramount, materials have often proved interchangeable. Over the past decade the competition between metals, ceramics, glasses, plastics, rubbers, concrete and timber has become fiercer, as the producers in one materials industry after another have increased capacity in excess of demand, and tried to boost their sales by price cutting and advertising.

The diagram 3.1 shows how the battle between various kinds of materials has proceeded over some years in the United States. Let us look at some of these materials in more detail. First, the plastics.

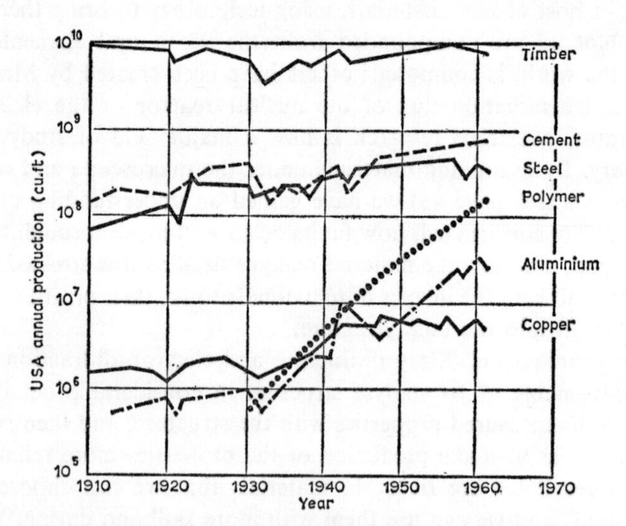

Diagram 3.1 Competition between the six major materials as seen in the United States market, by volume of production (Science Research Council)

The Plastics

The plastics are the most outstanding materials of the twentieth century, from which much of its technology has been coined. The first material of this type—cellulose nitrate or celluloid—was invented in 1868 as a substitute for ivory, but because of its inflammability and the difficulty of moulding it, its uses were restricted. Bakelite, patented in 1909, was the true basis of the synthetic plastics industry. Since then many more have been added to this new class of materials, which has had an astounding impact on the world. The world's use of plastics is expected to more than double between 1968 and 1975, when it should exceed 45,360 million kg (45 million tons). Table 3.1 (see page 62) shows some common types of plastics and their applications.

Of the two basic types, thermoplastics (such as polythene, PVC, polypropylene, polystyrene and polytetrafluoroethylene) can be softened by heat and restored to shape several times without changing their constitution. The thermosetting plastics, however, cannot be softened once set without being destroyed because they undergo a chemical change in setting. This class—to which Bakelite belongs—includes nylon, urea, melamine, epoxies, phenols, polyesters and rayon. There are other intermediate types like urethanes.

Chemically, they are all similar materials. Both classes are polymers, as the name usually denotes, made from 'building blocks' of other compounds called monomers. During manufacture of the plastics the blocks are linked together to form long chains and it is these long chains that give plastics their distinctive properties: they are soft enough to mould to shape, and yet can be set hard by heat, chemical processing, or radiation—or a combination of these. There are many different kinds of chains and substances that can be incorporated into them, and in this processing the blocks are interlinked. Often the chains are as well.

Because of the wide choice of blocks and chains and the additional substances that can be added to them, the plastics are extremely versatile. A plastics material can be made almost to a recipe to suit a given job, with the properties required. Plastics in general can be prepared quite simply and cheaply in reproducible form from raw materials that are themselves fairly cheap at present. They can be moulded into products, extruded into tubes and films, cast into sheets—which may then be formed into shaped articles—

Table 3.1

Common Plastics Materials and their Uses

Common Plastics Materials	*Uses*
Acrylonitrile-butadiene-styrene (ABS)	vehicle parts, pipe and pipe fittings, tool and utensil handles
Casein	buttons, buckles, toys, adhesives
Cellulosics	spectacle frames, toys, telephone hand sets, tool handles
Epoxy	adhesives, printed circuits, protective coatings for appliances
Nylon	tumblers, rope, brush bristles, slide fasteners
Melamine	tableware, buttons, laminated surfaces
Phenolics	electrical insulation, radio and TV cabinets, handles, dials, knobs
Polyesters	reinforced plastics for boat hulls, car bodies; impregnating mat or cloth
Polyethylene	squeeze-bottles and bags, flexible toys
Polypropylene	heat-sterilizable bottles, packaging sheet and film
Polystyrene	wall tiles, portable radio cases, refrigerator food compartments
Urea	electrical devices, lamp reflectors, buttons, stove knobs
Urethanes	foams for insulation, packaging, cushions; solids for tyre treads, bristles
Vinyls	gramophone records, floor and wall coverings, electric plugs, upholstery, raincoats

blow-moulded into bottles, foamed into microporous blocks, laminated to each other and to other materials entirely, and coated onto papers and fabrics. Diagram 3.2 illustrates a common machine for injection-moulding plastics: it makes plastics goods by squeezing a heated plastic charge into a mould or die.

Plastics therefore appear in an astonishing variety of shapes, sizes, colours, surface finishes and other properties; comprising the whole or part of cups, saucers, jugs and jars, brushes and dustpans, floor and other surface coverings, electrical switches and insulators,

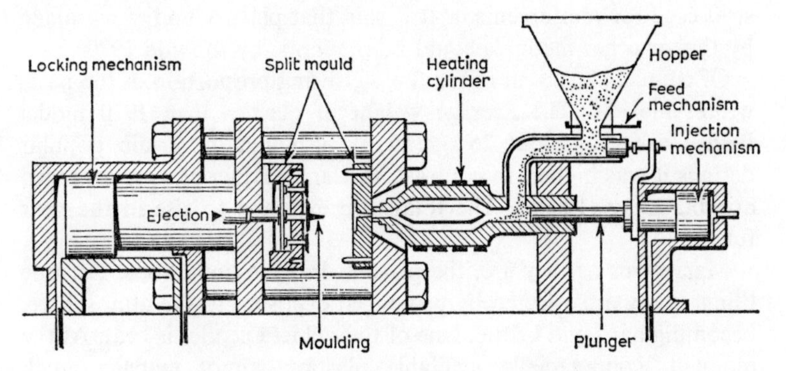

Diagram 3.2 A plastics injection-moulding machine

paints and adhesives, packaging, protective coatings, as well as clothing materials of all sorts, rivalling the natural fibres. As structural materials they emulate wood, bone, horn, ivory and—most important—steel and other metals, having advantages of strength and lightness and resistance to corrosion. The monomers for most of them are made from oil, via the oil refinery and the petro-chemicals plant. The notable exception is rayon, which is obtained from cellulose, one of the main constituents of plants, trees and cotton. Among the most vital raw materials from which plastics and synthetic fibres are made are ethylene and acetylene (colourless, inflammable gases) and naphtha, a yellowish liquid like petrol that is obtained in the distillation of crude oil.

Though they have broken into many lucrative mass markets, the plastics have not so far achieved much in the motor vehicle market, which could be the most lucrative of all. Experimental car bodies have been made in plastics for some years, especially of resin-reinforced glass fibre, but the breakthrough is yet to come, and the major motor manufacturers are keeping a wary eye on the relative prices of plastics (which are falling) and steel (which is rising). They will be prepared to change over when the price structure is right and the technical problems—such as the different way that plastics flow when stamped out in a press—have been overcome. At present, most of the plastics body shells are made from reinforced glass fibre which demands much hand labour and is not easy to fit into mass production. An alternative is to make the body shell in only a few pieces by injecting polyurethane foam into a mould containing a steel cage for reinforcement. It is said that plastics bodywork made by this or other methods could be economic by the mid 1970s.

Of course, plastics are used for a growing proportion of the parts within the car. The average weight of plastics in a 1970 model British car was about 25 kg (55 lb), although in certain popular designs it was higher. In contrast, the larger American car averaged around 45 kg (100 lb). The totals are expected to rise in the near future.

Marvellous as they are, the plastics do have limitations. For one thing, many are affected by prolonged exposure to the atmosphere, becoming hard and brittle. One of their chief enemies is heat. At the moment, commercially available plastics cannot manage much more than 260°–290°C for long periods, and 315°–430°C intermittently. Research workers trying to improve plastics' heat resistance have several options. One is to incorporate different substances—such as silicon, boron, and nitrogen—into the building blocks from which the polymers are made. The silicone polymers are notable, being inert to heat and also to chemical attack from many substances. They form a whole family similar to the naturally-occurring hydrocarbons that are vital constituents of all life. Among the members of that family are whitish greases that lubricate high-vacuum apparatus in the laboratory; silicone rubber tubes that replace diseased tissues in the body; silicone hydraulic fluids that operate controls in hot sections of the aircraft jet engine; barrier creams for the hands; and even astronauts' overshoes, in which they walked on the moon.

Technology is throwing up new variations of plastics all the time: the fluorocarbon series were developed as coolants, lubricants, sealants and buffer gases in plants making uranium for the American atomic bomb, and came to industrial notice through that programme. Most were discovered in the late 1930s. Great stability—in thermal and chemical terms—are their most useful attributes. The most familiar is probably polytetrafluoroethylene (PTFE), which forms washers, tapes, valve seals, piston rings in non-lubricated air compressors, sheathing for control cables and electrical insulation, and even non-stick linings for kitchen pots and pans.

The synthetic fibres have also enjoyed a great rise in popularity, as the graph (Diagram 3.3) shows. In Britain, the textile industry in 1968 used more man-made than natural fibres for the first time in its history, and the same trend is repeated across the world. The use

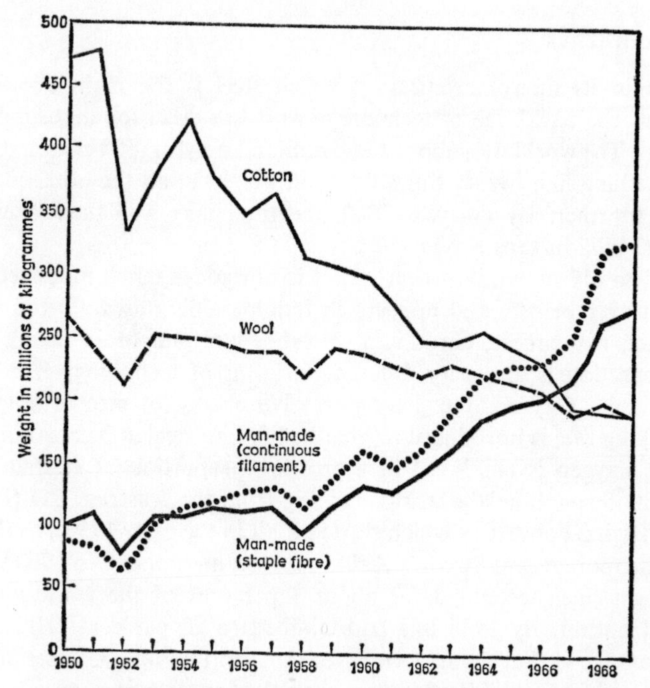

Diagram 3.3 Annual production of yarn in the United Kingdom from various sources, natural and man-made (Shirley Institute)

of truly synthetic fibres (excluding rayon) is expected to double to just under 7,000 million kg (15,000 million lb) a year in 1975 and to rise to 27,000 million kg (60,000 million lb) by the year 2000, helped by the non-woven and mixed fabrics. Over this time, it is anticipated, their share of the market will increase from nearly 20 per cent to 50 per cent. The world average use will then be about 7·7 kg per person (17 lb), although in the United States today it is already over 18 kg (40 lb).

This too represents a phenomenal growth. Rayon was only discovered in the latter half of the nineteenth century—independently in France and Britain. That was the first, followed by acetate, made, like rayon, from the natural cellulose of wood pulp or cotton. Nylon, the first truly synthetic fibre, made from chemical raw materials, was introduced by Du Pont in the United States in 1939. Since 1950 more fibres have been introduced in quick succession, among them the acrylics, and polyesters.

The Metals

Iron in its many varieties—of which steel is the main one—still forms the backbone of technology as it has done for thousands of years. The world uses about 5,000 million kg a year (500 million tons), something like twenty times the total weight of all the other metals put together. By the year 2000, the total may well have risen to 1,000 million tons a year.

If an old material at heart, steel is nowadays much more refined than ever before, and appears in innumerable guises alloyed with nickel, manganese, chromium, molybdenum and other metals. Its preparation is also very different in terms of technology if not in principle. Oxygen, the great corrosive enemy of steel during its working life, is introduced to breathe life into steel at its creation. In fact, oxygen is employed in a growing proportion of steelmaking plants throughout the world, ranging from the Austrian LD (Linz-Donawitz) converters which started it all in the early 1950s, to those of the more recent Swedish Kaldo process, and others. In 1960 basic oxygen steel accounted for about 4 per cent of the world's total steel output. By 1966 this had climbed to 22 per cent. The chief advantage is that the oxygen saves time. In the old processes an oil or gas flame was blown across a pool of molten iron in a hearth lined with refractory tiles (the open hearth); or air was blown through molten iron in a Bessemer converter, a large, upturned open-topped vessel. In an open-hearth furnace it took ten hours to make a 5,000 kg (500 ton) batch of steel. In the new basic oxygen furnace, however, where pure oxygen is blasted through a hollow lance into the molten iron at supersonic speeds, it burns out the carbon, phosphorus and other unwanted elements so quickly that the batch time has been cut to half an hour (Plate 3.1).

Even more radical processes are being considered. In one of them, spray steelmaking, a fine spray of molten iron is blown into a reaction chamber to mix with a spray of lime and a blast of oxygen. In a flash the impurities are burnt out, producing steel in a matter of seconds. In another, iron is leached out of its ore at relatively low temperature, then precipitated from the solution and finally reduced to metal powder in a stream of hydrogen or other gas. The powder, mixed with additives if necessary, is pressed or rolled into desired shapes directly, and the 'green' component only needs a final sintering procedure to convert it into the finished steel product. (Sintered metal products, important parts for such industries as motor vehicles, business machines, and domestic appliances, are usually formed from metal powders made in other ways, however.) This dispenses with the blast furnace altogether and saves much of the heat that is normally needed both to keep the iron molten (at around 1250° to 1450°C) in the blast furnace, and subsequently to re-heat the steel ingots before they are rolled to shape in the rolling mill.

Indeed, the whole steel industry is in ferment with new ideas for making iron and steel, and for forming them into products. Casting and melting under vacuum, for instance, excludes unwanted gases and prevents reactions with the air: hence vacuum-cast steels, free of hydrogen and other impurities, have improved properties and do not need a lengthy finishing heat treatment. Continuous-casting plants make ingots in much longer lengths than is possible with the normal casting in individual moulds, and confer the benefits of longer production runs, giving a more consistent product. The steel is continuously poured into a water-cooled mould in the shape of the required bar, or slab, or whatever it may be. This saves much time and expense in the rolling mill, where under conventional methods the raw ingot is rolled to its final form. The aluminium industry depends almost entirely on continuous casting, and the more hesitant steel industry is now accepting the process quite speedily. There are upwards of 130 continuous casting plants in operation in the world, and the number is increasing. Diagram 3.4 (see page 68) shows one type with a curved mould that brings the ingot out horizontally for cutting into lengths.

Many steelworks use X-ray and other analysis equipment to assess samples of metal taken from the melt to determine quickly—in some cases a matter of minutes—whether it has the proper chemical composition. Many use computers to control production and the sawing and cutting programmes so as to enable them to make the best use of the rolled length of steel to meet customers' orders.

With increasing international competition and growing mechanization, the steel industry is coalescing into fewer, larger units. At the end of 1967, the thirty-two largest companies in the world—excluding the Eastern block—were responsible for 70 per cent of total output: the twenty top companies claimed 60 per cent of the total. The same sort of pressures affect the steelworks itself. This is now visualized as a mighty, integrated plant, capable of producing 5

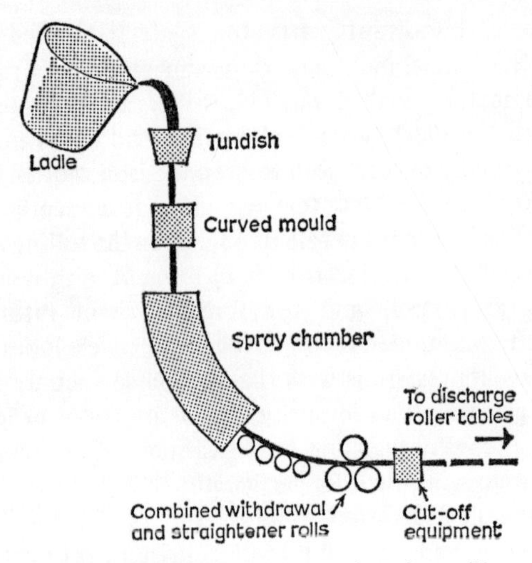

Diagram 3.4 Continuous-casting plant with a curved mould

million tons of steel a year. Built on the edge of a deepwater port where its iron-ore ships could discharge their cargoes, it would combine in a logical, efficient sequence all the functions necessary to transform the coke, iron ore, limestone and scrap arriving at one end into the girders, rods, tubes, sheet, coil and other semi-finished components that would emerge at the other. Such is the vision. A growing number of plants throughout the world are approaching that ideal.

If steel is and will remain technology's main prop, its supremacy continues to be challenged by other metals like aluminium. This versatile metal is one of the most widely distributed in the world, but not until the mid nineteenth century was a method of commercial production discovered. It began to assume industrial importance after the Second World War, and is now used in large amounts for cans, foil, cladding on buildings, heat exchangers, car radiator grilles, aircraft structures, and thousands of other purposes. Extensive use was made of aluminium in the American moon flights: in the lunar module and its foil wrappings, in astronauts' tools and space suits, and in the command/service module, which carried astronauts to the moon and back, aluminium was the primary metal.

Before the Second World War, steel and copper, lead, zinc and tin were the five main industrial metals. Since then a whole host of new metals has appeared. The really significant achievements in technology that started the new metals on their way to maturity were the gas-turbine engine and nuclear power which, together with their offshoots the supersonic aircraft and the nuclear-powered submarine, have made stringent demands on metallurgical technology. More recently, space exploration and the quest for even cheaper sources of power have intensified those demands. One generation's curiosities become accepted parts of the next generation's technology: for instance, uranium and plutonium are much in demand today as nuclear fuels. Zirconium found little commercial use until about fourteen years ago, but because it is incorporated in tubing and other parts of nuclear reactors, now faces growing demand. Titanium, which was not available commercially before the last war, claims an expanding market, where its lightness and strength at high temperatures make it an obvious choice for components in high-speed aircraft.

There are other curiosities, too, like the semiconductors, which are not quite metals and not quite insulators. Chief among them are silicon and germanium, but many more compounds are being investigated and tried out by the electronics industry (see page 56). The industry's needs for very pure and very perfect crystals of silicon, germanium, quartz, sapphire, etc, has in fact created a new technology centred around growing and processing crystals.

Chemical Technology

Through chemical technology, Man can adapt and improve on naturally occurring materials; he can make them more cheaply from alternative sources, with more consistent quality and higher standards of quality and performance; and he can formulate materials of all kinds with properties quite outside the scope of natural substances. Sophisticated, civilized Man desires not only the substances crucial for his technology as such, but also compounds to minister to his comfort and well-being—cosmetics, perfumes, soaps and detergents, medicines, and so on.

In antiquity chemical technology furnished dyes and bleaches for his textiles, glass, soap, metals, medicines, cement, ceramics, military combustibles and explosives, and other essentials. Since then its scope has broadened immeasurably to embrace a vast array of substances from fertilizers to plastics, perfumes to explosives, drugs to weedkillers, paints to adhesives.

The chemical industry is thus of prime importance to all technology, to Man's comfort, and his vanity. In recent years it has been expanding at a great rate—about 10 per cent a year in Europe and the United States, or about twice as fast as manufacturing industry overall—and its rate of investment and productivity are higher than for industry in general. It is what we may call a high-technology industry. In the United States its growth has been especially remarkable, and from the time of the Second World War the American chemical industry forged ahead of those in other countries, increasing its output fourfold between 1939 and 1950.

The chemical industry's modern growth began with textiles, as the textile dyers needed great quantities of alkalis (carbonates and hydroxides of sodium and potassium), acids (sulphuric, nitric and hydrochloric), and chlorine compounds for cleansing, bleaching and dyeing the fabrics. At the beginning of the nineteenth century the production of these compounds, together with the smelting of metals, were the main chemical trades. In spite of its enormous expansion, these same acids and alkalis are still among the basic substances required in heavy quantities from the chemical industry. The use of sulphuric acid, for instance, or of steel, or electricity, is often taken as a measure of the well-being of a technological economy. This vital chemical, one of the first to be made on a large scale, is employed in the manufacture of fertilizers (chiefly superphosphates and ammonium phosphates), titanium dioxide (a white pigment introduced in a variety of goods, from paper to cosmetics), soap and synthetic detergents, dyestuffs, petroleum products, and many more besides.

Similarly, the textile makers' demand for new dyes led to the establishment of the organic chemical industry, which was virtually non-existent before 1850. The chemical plant and the intermediate compounds involved in making dyes are just as useful for making synthetic drugs, perfumes, solvents, fine chemicals, and explosives, and these were soon exploited. Although the industries based on these products were almost exclusively in German hands from the start, around 1880, the occurrence of two world wars in the first half of the twentieth century forced other countries to develop their own organic and inorganic chemical industries in the face of the German monopoly.

In particular, the mass-production of the motor car demanded massive quantities of rubber for tyres and steel for many kinds of parts. When supplies of natural rubber from Malaya and Indonesia were cut off in the Second World War, a crash research and development programme was initiated to find a general-purpose synthetic rubber—ironically with the aid of pre-war German research. From then on the production of synthetic rubber climbed, until in 1959 it equalled that of natural rubber. A vast new industry had come into being.

At the same time, the large-scale refining of petroleum for fuel for cars also created substantial stocks of by-products which came to be used in the petro-chemicals industry, for making both existing products and new ones, synthetic rubber among them. Since then chemical producers have come to rely more heavily on petroleum for their raw materials where they once relied on coal tar or agricultural products. For example, in 1950 only 9 per cent of organic chemicals produced in Britain were based on petroleum; by 1965 this had risen to over 70 per cent.

One consequence was that petroleum-derived primary chemicals were available on a large scale for detergent manufacture. It was just as well, with a growing world population clamouring for higher standards of cleanliness and better ways of reducing the tedium of the ancient chore of washing. In the earliest days people applied oils or greases to the skin, scraping off the excess and the dirt with it. Then soaps were made from alkali and palm oil, tallow, or other natural oils, and though adequate for a time, they were not good enough for a technological society whose hands were begrimed with the mineral oils from its machines.

Synthetic detergents were brought into use around 1920, the industries being rapidly fostered during the last war in Germany, the United States and Britain. The sludge left after the treatment of petroleum compounds with sulphuric acid to remove sulphur, oxygen, and nitrogen compounds provided a cheap source of the necessary intermediates. The UK production of synthetic detergents jumped from almost nothing just before the last war to 215,000 tons in 1953, 350,000 tons in 1965, and 484,000 tons in 1969.

Continuing to improve on Nature, modern manufacturers usually ensure that their detergents contain 'buffers' to keep the solution alkaline, polyphosphates to counteract any hardness in the water, and optical bleaches that emit blue light to correct the slightly yellowish tinge that clothes normally acquire with age. The latest types are biologically active, containing in addition enzymes which 'digest' organic dirt instead of merely dissolving it. These enzymes—living catalysts—and the technology that creates them, provide good examples of the use of biologically active substances, whose ability to bring about subtle chemical changes in the most unlikely substances is exploited in widely differing fields. These range from fermenting beer and wine, and breadmaking and cheesemaking, to making antibiotics such as penicillin, and at the other end of the technological cycle, converting noxious waste materials into harmless ones at the sewage works and in new kinds of plant that turn refuse into compost (see page 273).

It is rare for major chemicals to be made today without a catalyst coming into play at some stage or other of the synthesis, and such basic chemicals as nitric acid, methyl alcohol, phthalic acid and formaldehyde are mainly manufactured by catalytic processes. In fact, catalysts may be said to be among the most important classes of materials in the world, as they provide the trigger for a whole range of chemical reactions, from the combination of hydrogen and nitrogen under high pressure and temperature to form ammonia, to the 'cracking' of petroleum in the oil refinery to yield more useful products, and the hydrogenation of groundnut oil to produce one of the ingredients of margarine. The reactions either would not happen without the catalysts, or would be so slow as to be industrially useless. Though the basic principles of catalysis were spelt out in the nineteenth century, commercial exploitation has been a feature of the technology of the twentieth. It is a constant challenge to the research chemist and metallurgist to produce more efficient and cheaper varieties; in the early 1950s there were some rare successes, such as the Ziegler catalysts, which opened a route to make relatively high-density polyethylene plastics at normal temperatures and pressures, by-passing a high-pressure process producing low-density polyethylene that had been used for over twenty years. Indeed, the large-scale production of plastics relies on the ability to make their raw materials in large volumes at low cost, which in turn relies on catalyst development.

Each chemical treatment is a series of step processes, such as hydrolysis, nitration, oxidation, etc., and each process in turn is a series of unit operations—mixing, heating, precipitation, evaporation, etc. These apply whatever the scale involved. Close study of each of these steps on theoretical and practical bases and especially by the rules of thermodynamics, has enabled the chemical engineer to design and build chemical plant in a much shorter time, and with a greater chance of its doing what it is supposed to do, than was possible with rule of thumb. Each step is under continual improvement.

Another trend is apparent—that of changing from batch production to continuous production, wherever possible, and of eliminating expensive labour and hence introducing a greater degree of automation. Until 1920 the chemical industry used batch production methods and depended on equipment makers for the merely minor improvements in an essentially stable technology. Then the change to continuous production methods created a need for new technologies that equipment makers could not supply, and independent engineering firms stepped in, working with the chemical companies and equipment makers. The latest improvement has been to introduce comprehensive instrumentation and control equipment, usually connected to a central control room supervised by a few people, to keep watch on pressures and temperatures, flow rates, and other vital indicators of the health of the plant. A plant producing perhaps hundreds of thousands of tons of chemicals a year can thus be controlled by only a few people (see Plate 3.2). Ultimately, it will be controlled by computer: already some sizable refineries and chemical plants are coming under computer control.

In this field, as in the mass production of goods, a large scale of operation reduces costs in relation to the unit of output. Because of that, the scale of chemical plant has been steadily increasing. Take the case of the production of ammonia, a raw material for fertilizers, and an intermediate in making nitric acid. Only a few years ago a plant making 10,000 tons of ammonia a year was regarded as quite sizable, and yet three new plants built by Imperial Chemical Industries in Britain produce 300,000 tons a year each. Their combined output is enough to satisfy the demands of the entire British agricultural industry, and more. The manpower required—which is an inverse measure of their technological sophistication—amounts to only 0·3 man-hours per ton of ammonia, as against 11·6 man-hours per ton needed on older plants. Nevertheless, there is still plenty of room for the small, specialist company with only modest production facilities, as far as size goes.

Most products of the chemical industry are made at high temperatures because most reactions run more swiftly the higher the temperature. However, certain products are made by low-temperature processing—the outcome of studies on air liquefaction carried out at the end of the nineteenth century. The existence of liquefied oxygen, nitrogen, hydrogen and other gases in substantial quantities has enabled the new science and technology of cryogenics (studies at very low temperatures) to become something more than a laboratory technique.

Until recently, when the demands of space rocketry spurred the demand for liquefied gases as fuels, cryogenics was a restricted field. There were few means of creating low temperatures, and these did not extend much below −70°C. Indeed, for one reason or another, Man seems to have neglected the study of cold and its effects in favour of the study of heat and what it can do.

But discoveries with gas liquefaction from the late nineteenth century on removed the technical barrier and enabled low temperatures to be attained easily through compressing and liquefying such gases as oxygen, nitrogen and helium. The great impetus came from the astronautics business, which needed liquid oxygen, hydrogen and other gases, as fuels. Steelmakers, too, were able to benefit. The iron and steel industry now devours vast quantities of oxygen and nitrogen especially. As a matter of routine, industry can order tonnage quantities of these gases, and of rarer kinds, such as argon (used for filling lamp bulbs and shielding welding arcs to prevent oxidation of the metal), and xenon (for filling high-power light bulbs), helium, much used for research, and many others. Growing quantities of nitrogen are used in freezing food.

Among the effects produced at low temperatures, none seems to be of more interest than superconduction. At extremely low temperatures—typically 20° or less above absolute zero—certain metals and alloys lose their electrical resistance: they become superconducting. A current started in a superconducting loop of wire will flow indefinitely. Because there is no resistance, the current generates no heat, in contrast to the troublesome behaviour of normal conductors. Thus very heavy currents can be passed through a superconductor, generating very high magnetic fields around it.

This effect should have applications in wide fields: in exceptionally powerful magnets, used in the laboratory to examine electronic and atomic characteristics of materials; in superconducting electric cables that would distribute heavy currents without the heat penalties associated with normal cables; in radio telecommunication receivers, where the unwanted background noise can be removed by cooling the receiver device in liquid helium; in lighter, smaller electrical motors and generators which, in theory at least, could propel or turn a great variety of machines, trains and ships among them.

Ceramics, Composites, and Others

Modern technologies, and the nuclear and aerospace industries in particular, need materials capable of awe-inspiring feats of endurance, withstanding heat, mechanical stresses, corrosion attack, radiation and any other assaults that Man and Nature impose. One of the first solid materials that Man used was a ceramic, stone. Later he made his own ceramics—pottery and bricks, glass and concrete. Now, thousands of years later, major industries revolve around these products. As it turns out, certain types of ceramics can be applied no less successfully to the latest demands of technology—in linings and other structures that withstand formidable temperatures in rocket motors, jet engines, steel-making furnaces, and elsewhere. Their magnetic properties are also used in such items as the ferrite cores that store information magnetically in computer 'memories'. Still at the research stage is an electrically-controlled ceramic plate that turns white light shone through it into coloured pictures, and could be used as an optical display system, perhaps a substitute for the cathode ray tube (see page 52).

The field of modern ceramics stretches far and wide; some of the better known varieties—such as ruby and sapphire, porcelain, cement and sand—are being supplemented by compounds like the carbides and nitrides (that do not occur in Nature), thanks to the better understanding of structure and properties afforded by materials science. Carbides, such as tungsten carbide, and oxide ceramics are commonly replacing steel tools for cutting and shaping metals, and carbides are extensively used also in punches and dies employed in the manufacture of items like razor blades. Versatility, both in the type of ceramic and the ways in which it can be fabricated, is one of the most valuable properties of these materials.

'Glass' is also a versatile substance, and the one name, like 'ceramics', covers many variants. Though the principal ingredient is silica sand, almost all elements can be added in various proportions to endow it with different properties. More than a thousand kinds of glass are made commercially, including such diverse products as glass fibre strand one-tenth the thickness of a human hair, in which a modulated laser light beam can carry thousands of telephone conversations (see page 54), to the humble milk bottle. The main kinds are the ordinary soda-lime type—for windows, bottles, etc.; lead-alkali varieties, for optical parts, tableware, etc.; borosilicates, for cooking utensils, laboratory apparatus; and the pure, or almost pure, silica used for high-temperature laboratory apparatus, spacecraft windows, telescope mirrors, etc. One of the most intriguing modern variants are the glass-ceramics, which are first made as a glass, then partially converted to ceramic by a carefully controlled heating process. About four times as strong as glass, they may be used, for instance, in the nose cones of radar-guided ground-to-air missiles, cooking utensils and stove tops, and in deep-diving vessels.

However, it is in the field of composite materials that the most exciting work of materials technology is going on. The composite contains thin, strong, brittle strands of one material in a bed, or matrix, of a more pliable one. There are many natural composites. Wood, for instance, consists of cellulose fibres in a matrix of lignin. The bamboo stem, which bends without finally breaking even after being deeply scored with a knife, provides a good example of the tenacity of the natural composite. Yet man-made composites can surpass natural ones in those properties that the structural designer values: they are strong, stiff, light, resistant to shock and to 'creep' under prolonged load. Composites, in fact, can be lighter, stiffer and stronger than anything previously made.

The implications for structural design are profound. At present composites are still too expensive and too scarce to be used much outside the fields of aerospace—which spurred their development in the first place—and deep-sea diving. In due course, though, they will surely be put to many more purposes, such as the construction of buildings, boats, golf clubs, skis, racing car bodies, and so on.

There are innumerable variations that can be played on the basic composite theme. Fibres can be long, short, or even continuous; thick (up to say, 0·025 cm in diameter) or thin. Many materials—including glass, metals, ceramics, carbon and boron—can be obtained in fibre form. Normally a fibre is prepared from the bulk material and with the exception of asbestos can usually be continuous. If it is grown on the parent material from a gaseous compound it tends to form in short, bristle-like single crystals, called whiskers. Some 'fibres' can be formed by careful treatment within the body of some alloys. The matrix may also be very much varied. It may consist of only one component, such as a resin or elastomer, a metal or ceramic. It may have several components, and may deliberately include holes, to save weight, perhaps, or to hold a lubricant or hard particles that make it wear-resistant. It may have some special structure, such as aluminium alloy honeycomb that has been resin-bonded to sheets of metal or other substances.

Resins reinforced with glass fibre are probably the most widely used composites—discounting for the moment substances like concrete, which could be regarded as a special type of composite. One point about glass fibre is that it can be obtained in very long filaments from which large vessels—for holding gases at high pressure, say—can be wound as cotton is wound into shape on a bobbin. But glass fibre is not very stiff, and tends to weaken at temperatures much in excess of 400°C. Fibres of carbon and boron, however, are much stiffer and more resistant to high temperatures. These two properties, combined with their high strength, mark them out for wide application in the future, though their price is still high. Their early uses include compressor blades in aircraft jet engines, helicopter rotors, parts of aircraft fuselages, and components such as gearwheels and bearings, where the carbon fibres' lubricating properties are valuable.

Resources for the Future

Thus technology now enables Man to extract and utilize materials from his environment on a scale unthinkable a hundred years ago. It may be a truism to say that technological advance depends on new materials, but it is well worth remembering. Structures are only as good as the materials they are made from. Similarly, the development of new processes has always had to go hand in hand with that of materials. Many of the alloys in use today have no real worth beyond the shaping processes for which they were designed—for example, zinc alloys for pressure diecasting for chrome-plated car lamps.

Materials development, and with it technology, has taken great strides forward in the last thirty years or so, generating in the human race a feeling of unequalled prowess, a confidence in its technological ability. There is no doubt, of course, that improvements in materials did a great deal to free technology from the constraints imposed by natural materials. Modern substances enable Man to design and construct his vast technological infrastructure—buildings, vehicles, machines, implements, and so forth—with a freedom he has never achieved before. In ancient times the designer simply did not conceive of things that required anything more than stone, cement, clay, the alloys of the seven ancient metals, and the common biological materials—wood, horn, etc. One of the great moments in Man's history came when he first discovered that he could alter the nature of materials, changing clay to stone by carefully heating it for instance. For early Man, that must have seemed an almost godlike power. Modern Man still feels a touch of it. Each of the vast number of other transmutations that he has attempted and achieved is a cause for self-congratulation, a feeling of satisfaction that another corner of Nature's kingdom has fallen to his power.

While technology enables Man to extract materials from Nature's grip, they are not his for long. Once separated from the environment and converted into a useful form, they are open to attack by all the environment's forces trying to win them back again. Wind and rain, often laden with pollutants or salt spray, corrode them; friction wears them away; dust scours and buries them; sunlight ruptures their chemical bonds; micro-organisms seek to downgrade them.

Man displays considerable skill in protecting materials from these scavengers of the environment and preserving them in the form that he wants. He may do this by preparing alloys (such as stainless steels) that do not rust or corrode except under extreme provocation; by coating materials with paint, zinc, plastics, or other resistant sheath; by an electrical technique, at the expense of a 'sacrificial' electrode, and in many other ways. In time, however, Nature wins.

While it lasts, the rapid advance of materials development lulls us into a false sense of well-being and confidence in the technological mastery of our environment. But it cannot last long, in terms of human evolution. So heavily have the earth's mineral reserves been tapped that in the past fifty years more has been extracted from them than in the entire span of previous history. Technology has a voracious appetite. With the growing world population and the rising use of materials per person, the rate of extraction has been increasing still more rapidly in recent years.

As the use of plastics and synthetic fibres has risen—doing so rapidly since the last war—so has the demand for raw materials from which to make them. The introduction of nylon, polyethylene and other products on their vast scale has made the earlier sources of raw materials such as coal and coal tar inadequate in quantity and quite uneconomic. But the petrochemical industry has been able to step in to fill the gap with raw materials derived from oil. That was an easy enough switch.

But still, our reserves cannot last indefinitely. Indeed, some of them will be feeling the strain fairly soon. Even at current rates of use, the known reserves of most staple minerals will be exhausted within 180 years, many of them much sooner. Among the metals, gold, zinc, lead and platinum are in danger of depletion before the year 2000. The currently commercial reserves of silver, tin and uranium will also be in danger by then. So will those of oil and natural gas. Resources of coal and peat are dwindling, and are unlikely to last for more than a few centuries. Since we rely on these hydrocarbons for both fuels and materials, via the petrochemicals, they are vital to our present level of technological activity. When they disappear our industrial life will be in jeopardy. The discovery of a new large oil field in North Alaska in 1968 was one of the increasingly rare finds; there will, I expect, be others, in remote regions and on the continental shelves beneath the oceans most probably, but they can do no more than buy time. Mineral resources, oil included, are the products of millions of years of the earth's history. They are not renewable, in so far as new deposits do not form fast enough to replace those that we extract. Mines, unlike fields, do not go on producing.

Faced with the prospect of dwindling resources, technology must be applied on a vast scale to tackle a great array of problems concerned with the basic themes of discovering new deposits of mineral resources other than those already known, mining or otherwise extracting lower grades of ore; devising better substitute materials for those in especially short supply; and in particular, finding improved methods of recycling used materials—to imitate Nature more successfully, with some modifications. As to hydrocarbon reserves, we shall be forced back on to the self-renewing reserves that can provide the same carbon-containing compounds that we need—that is, plants and perhaps animals. (Nuclear power will give us some respite, see page 134.)

This will spur exploration and exploitation in many hitherto untouched areas, like Siberia and Alaska, on land and at sea. Nonetheless, because the materials will have to be won from remote places and brought to their markets, they will be more costly, and it will be increasingly preferable to recycle materials from wastes instead of getting them from virgin reserves.

There are much more serious problems looming than those of price, however. As industrial nations use up their own accessible supplies they have to import more from other states—generally those that are still developing their industrial economies. For example, at the beginning of this century it was believed that the Great Lakes iron ore reserves in the United States were unlimited, but now, only seventy years later, they are diminishing rapidly and the United States has to bring in iron ore from Canada, Venezuela and elsewhere to meet its massive demands. All industrial nations, except perhaps the Soviet Union, are net importers of most of the metals and ores on which their economies depend. They obtain these supplies from developing countries that do not possess the technology to make use of the materials themselves. For how long will the developing countries be prepared to export their raw materials in this way, buying back goods of very much higher price? By so doing they may be denying themselves the means of future industrialization, while engaging in a trade that is weighted against them. So far, the political and economic consequences of the growing competition for materials have been mild. As accessible supplies vanish, though, the competition will become fiercer and its economic and political repercussions more violent.

4 · The Versatile Computer

Birth and Growth

The electronic computer would be technology's most successful machine were it not for the difficulty that people have in accepting it. Speed is its essential quality. Working perhaps 400 million times faster than a man, it can perform in a couple of seconds all the arithmetic that he could do in a lifetime. It can collect, process, store and retrieve all kinds of information in a minute fraction of the time it would take by manual methods—irrespective of whether the information concerns people's tax codings in government records, spare items of machinery in a factory stores, or atmospheric weather patterns.

Previous machines supplemented Man's physical powers, enhancing his ability to shape and use materials and relieving him of much unpleasant physical work. The computer, which contains the electronic equivalents of human logic and memory, supplements his mental powers, enhancing his ability to 'shape'—or process—information and relieving him of mental effort. It can free people from a tremendous amount of routine sorting, tedious calculation, filing and book-keeping, if it is used properly. But these are early days for the computer as yet, and few people really know how it should be used. For every person who finds it an exhilarating experience to work with a computer, there are probably two or three who find it unnerving. The early, incorrect nickname 'the electronic brain' created an atmosphere of awe and mystery which still has not been dispelled, to everyone's disadvantage.

The comparison with the brain is at present far-fetched. True, the computer is many times faster, but it cannot match the sheer capacity of the human brain, with its 10,000 million nerve cells, crammed into every corner of the cranium with their connecting nerve fibres and insulation. Each cell may be connected to 10,000 adjacent cells, and so may involve them in its own activity. However, even the most complex present-day computer is unlikely to have more than a million units equivalent to nerve cells, and each one is connected in only five or six different ways. Most important, the computer does not have an imagination, has no desire to compute (or not to compute), and possesses no inkling of its role in relation to Man. All that it does possess must be put there by a man first of all.

The digital electronic computer was born in the 1940s, in the closing years of the Second World War. As might be expected, there was a military motive behind it. Similar machines were evolved in three countries—the United States, Germany and Britain—more or less simultaneously. Incidentally, the mechanical forerunner of the electronic computer appeared in Britain as long ago as 1832, conceived in the mind of Charles Babbage, who from 1828 to 1839 was Lucasian Professor of Mathematics at Cambridge. He started to make the small 'difference engine', as he called it, and succeeded in building various parts for it; but he never finished the whole machine. His much more ambitious 'analytical engine', which could have done all that a modern general-purpose computer can do, was conceived twenty years later. That was never finished, mainly because Babbage was never satisfied with what he had made, and continually tried to improve it. Nor was the technology ready. In the end his work was forgotten, and only with the coming of the thermionic valve and electronics did the computer—electronic and not mechanical—become feasible.

Since the war, in the span of one human generation, the computer industry has spawned three distinct generations of its machines (the first generation corresponds, roughly, with the use of valves, the second transistors, and the third integrated circuits—see page 48). In that short time it has become one of the most important industries in the world. So great is its potential in the handling of information that it is expected to become the world's largest industry before all that long.

There were about 400 computer installations in the world in 1953, and 60,000 in 1968; and it is estimated that by the mid 1970s there will be 175,000. In this vast growth the United States has far outstripped its competitors, both in terms of making and using computers.

The industry is growing worldwide at rates of 40 per cent and more a year, and in some places by as much as 60 per cent a year. But the pace is gruelling. To be able to stay in the business at all, firms have to be large enough to find the heavy investment required to search continually for new machines and better ways of using them. It can well cost £15 million to do the research and development for a large data-processing computer, with its ancillary equipment and services.

Production might start about four years from the initiation of the research programme, so that the firm could not start to recoup its investment for five years at least. Even then, so fast is technological evolution in the electronics and computer industries (and it is the pace in electronics that largely determines how fast computer development goes) that the computer model is unlikely to have a commercial life of much more than four years. Hence the firm would have to spend about £5 million a year just to do enough research and development to produce a new range of machines every four or five years. After that, it would have to sell enough machines to support the high level of R. and D. expenditure. This explains why it is so difficult to make computer companies viable, and why most rely on government help in the form of direct contracts for machines (e.g. for central and local government, defence agencies, and so on) or grants for R. and D., direct investment, and in other ways.

The U.S. computer companies—headed by International Business Machines Corporation—took an early lead in the industry and have since consolidated their position to the consternation of the rest of the world. Worried by this domination in such a vital industry for the future of technology, the Soviet Union and China (which have gone their own ways right from the start), Europe and Japan are all trying to build up viable computer technologies to match the American challenge. Even so, a firm in the United States may very likely be large enough to spend twice as much as any European firm, have twice as many people working on a new computer, and produce a comparable machine in three years instead of four. That year of 'lead time' may easily give it the chance to snap up orders which a European firm would need to get in order to cover its R. and D. costs.

Anatomy of an Installation

Although there are two distinct types of computer, analogue and digital, the digital type is by far the commoner. When the term computer is used here, it means the digital type unless otherwise specified. The analogue, which simulates the behaviour of some kind of system, is used chiefly in research or design work. It is finding growing employment, often coupled with a digital machine, in controlling chemical and other processes, but it is overshadowed by the digital machine. As its name implies, the digital machine deals with digits—whole numbers—with which it performs the usual mathematical processes of adding, subtracting, and multiplying. However, this means that all quantities it deals with must first be converted into numbers, and this requirement is at the root of many problems.

A computer installation is basically in two parts: hardware and software. The hardware comprises everything that can be seen and touched—the computer itself, the control console, the magnetic tape cabinets, discs, and other forms of storage of information, and the various types of peripheral device that feed the computer with information and take information out of it—including paper tape readers and punches, card punches and readers, high-speed printers, cathode-ray tubes which perform the same function visually, and remote terminals such as teletypes or keyboard displays. The illustration (Plate 4.1) shows a large business installation.

The software is the collection of instructions and know-how that is needed to run the computer and its peripheral machines. The most important part of the software is the programs, or sets of instructions that tell this mass of machinery how to work—whether, for example, to work out a firm's payroll, to arrange the production schedule in a steel mill, to do a meteorological forecast, to control a city's traffic, to translate Russian into English, or to do any of the 1001 other tasks the computer is capable of. At present, the costs of hardware and software are more or less equal. But while the cost of hardware, thanks to some neat technical tricks, is rapidly decreasing, that of software is not falling so fast. In the future, a greater proportion of the cost of the installation will be for software—say 70 per cent of the total by 1975. This is because, generally speaking, a new program has to be written for each task the computer is called upon to do, unless it is a very minor one. It is true that there are expanding libraries of programs from which a user can take one to do a specific task which is common to a variety of users; but even then he will probably have to make at least some alterations for his own particular job, since that job, although it may have many factors in common with a great number of others, will nevertheless be unique, demanding a unique program. The expense is considerable. One programmer at present costs the firm employing him (or her) about £6000 a year a year, including salary, the firm's overheads, and the cost of the time he takes testing his programs on the computer.

Over the last ten years the speeds of the fastest computers have risen from 2000 to 500 million operations a second. At the same time the cost to a typical user of performing one operation—say, making a million additions of two ten-figure numbers—has fallen from £5 to 12·5p. This trend is likely to continue, aided by advances in electronics that produce cheaper, faster circuits for the computer's logic and memory. This will bring the cost of computers down and put them within the reach of many more users of all kinds.

The computer user does not have to buy one. Most are rented or hired. Nor does the machine actually have to be in the building. The user can rent computer time, transmitting data to it over telephone or other lines and receiving results back again over the same pathway, or even sending data in the form of punched tape, or in statistical form. No one need complain of a lack of facilities. Time-sharing is the latest version of computer service. In this method, a number of users are connected to the computer by high-speed data transmission lines, or public telephone lines. Each user has a terminal of some sort on his premises. The computer attends to all the users at once: at least, that is how it appears, and each customer has the illusion that he is the only one using the machine. Systems with fifty consoles

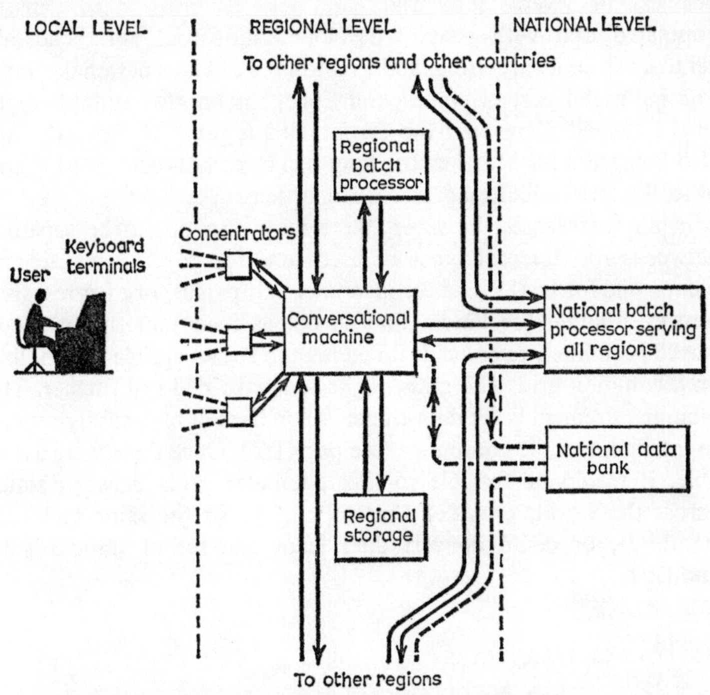

Diagram 4.1 A typical computer time-sharing network of the near future

or so in simultaneous use are common: larger ones may have as many as 200. The number is increasing steadily. This type of business, which began in the United States around 1964, is now well established there and is beginning to catch on in Britain.

With the coming of cheap computing power promised by the pundits, every home is assured of its own computer, or terminal. It is believed that one home in a hundred in the United States and Britain will have a computer within the next fifteen to twenty years. It would be able to carry out a variety of jobs for the owner. It could prepare his personalized newspaper, containing only the items in which he was interested, make out a list of his engagements for the day, order goods from local shops, and pay the bills for household expenses. The home computer or terminal would be linked to the local time-sharing network and the local centre of this network to the regional centre, which in turn would be linked to the national centre, perhaps with more levels in the hierarchy (Diagram 4.1).

Nascent computer grids of this kind are in evidence in research and business in several industrial countries. Britain—small, densely populated and with heavily developed industrial and academic centres—is an ideal setting, and London the obvious national centre. The industrial part of continental Europe is another suitable area. In the United States and the Soviet Union regional networks are also being established, but the long distances between centres, and other factors, militate against national networks.

In all these cases, however, there may turn out to be separate networks for different types of user—one for government departments, one for banks, one for insurance companies, one for research centres, another for hotels, and so on, as their patterns of needs would be unlikely to coincide far enough to be susceptible to handling on a common grid. The grids might be subdivided still further. The main requirement is for data-transmission lines of high enough speed to interconnect the computers (see page 156). Once these were available, it would be feasible to link computer grids between states across the world, provided that they observed the same technical standards, or could convert data from one set of standards to another.

Aspects of Hardware

There have been significant changes in the machinery of the computer installation—the hardware. The use of transistors, and more recently integrated circuits, has enabled computers to become smaller, faster in operation and cheaper in terms of the number of operations they can perform for a given cost. There have also been radical improvements in reliability. The first computers of the mid 1940s seem impossibly crude by comparison: one of the earliest machines occupied a whole room, consumed great quantities of electrical power to drive its thousands of valves, and broke down frequently. The integrated circuit can replace hundreds of separate components on printed circuit boards, with their many soldered joints, each of which is a potential source of unreliability (see page 48). In a typical computer with, say, 100,000 individual transistors, the saving in wiring and soldered joints achieved through using integrated circuits is tremendous.

Their speed has also been of great value. Since the computer is essentially a collection of switches made up from electrical circuits, the faster those switches operate the faster is the computer as a whole. Valves, used in the first-generation computers, can switch in a millionth of a second—a microsecond. (For comparison, it takes a good part of a second to switch off an electric light by hand, and about a thousandth part of a second—a millisecond—to do it with an electrical relay.) Semiconductor devices have improved on that by a factor of at least 100. But users are never satisfied with computers any more than any other product, and when some improvement is made, they always want something better. Thus, if the speed of the computer's electronic elements has risen a million times in twenty-five years, new 'real-time' tasks such as pattern recognition, missile tracking, and control of terrain-following radar make even greater demands. The search for faster operation is intense. Design of the computer layout plays an important part in this search, and the fastest computers of the future may well handle calculations in parallel instead of in sequence, as at present: that is, the computer will work on calculations simultaneously, instead of having to wait until it has finished one before it can start the next. However, faster electronic elements are vital, and all kinds of technical means are being adopted to reduce their switching speeds towards the nanosecond—a thousand-millionth of a second—and beyond.

A computer more than 0·3 m (1 foot) across could never work in a nanosecond, though, as it would simply be too large: the electric signals take about one nanosecond to travel that distance, and a correspondingly longer time to travel any further. While microelectronics has done wonders in 'compressing' great numbers of computer elements into small space, there are limits to the possible reduction in computer size. The smaller the computer, the more likely is interference between various components, and the more difficult it is to remove heat generated by the passage of electric currents. Although cryogenics should be able to alleviate these problems, there has to be a compromise between the desire for speed and the limitations of miniaturization.

Theoretically, an optical computer, using light instead of electric currents, would be faster. The fastest processes known to physicists—apart from certain subnuclear reactions—are interactions between light and atoms or molecules, and it may prove possible to make optical computer elements that switch in a picosecond—a thousandth of a nanosecond. This appears to represent the ultimate speed attainable by exploiting natural phenomena.

The computer's memory elements must also act swiftly, and delays caused by interconnecting wiring be minimized if the advantage of the rapid switching—or 'logic'—elements is not to be lost. At the moment, the fast main memory is usually built up from arrays of ferrite cores in which data are stored magnetically, strung on criss-crossing wires, but there is a limit to the amount of information they can hold if they are to be fast enough. To back them up, slower, larger stores, usually in the form of magnetic tapes, drums, discs or cards (equivalent to document files in the office), retain larger quantities of data from which the central computer unit selects portions at a time for processing. Again, there has to be a compromise between making a store large and slow, or small and fast. One line of attack is to increase the density of storing the information. Many new forms of memory element are being investigated, including those built up from integrated circuits; wires and thin metal films in which data are stored magnetically; and optical types, where a laser can be used to enter and extract information on film of some kind. There are many variants.

Successive reductions in the size of electronic circuits, and increases in the number of components that can be made on a single silicon chip have brought to the stage of feasibility computers on one, or a few, chips; indeed hundreds of components can be compressed into a chip only about 0·3 cm square to perform all the arithmetical functions of a medium-size, medium-speed computer of conventional design (see page 52). There is no doubt that further spectacular advances are in store. The latest type of LSI circuits offers computer logic elements at a price so low that a computer will be able to incorporate spare circuits, already wired up to the rest of the machine, which would be called into use when required by an automatic testing routine so that the computer could continue to repair itself for as long as its spare part bank lasted. Such a machine is being designed in the United States to control the unmanned spacecraft that will make the 'Grand Tour' of the solar system in the late 1970s. The tour will last from eight to twelve years, and so the computer will have automatically to locate, diagnose and repair faults inside itself and in other equipment during that time (see page 193).

However good the computer, though, it is still helpless without the peripheral devices that feed it with information and take the results of its labours for presentation to the human world. The basic problem is to make them fast enough. Punched cards are convenient data stores for commercial purposes, but handling techniques are slow, with card readers taking, say, 1000 cards a minute, and card punches 250 a minute. Current photo-electric paper tape readers work at something in the region of 1000 characters a second, but someone has to prepare the punched paper tape in the first place, and as yet the human eye and brain are unrivalled for their ability to recognize badly-written letters and figures quickly. Much work is going on into methods of optical character recognition, and some machines are able to take over that function with limited success. Nonetheless, before very long successful optical character recognition devices will be able to read documents of all kinds directly, and transform them into the sort of signals that the computer can process, so eliminating the toil of teams of girls who have to punch the information on to paper tape or cards at the moment. The mechanical printers that produce readable documents from the computer's output are amazingly fast in human terms, being capable of producing 2000 lines of text a minute, but this is too slow for the computer. One way of accelerating the process is to feed the results from the computer into an electrostatic printer, which can operate at rates of 82,000 words a minute and more. Another is to feed them into a special cathode-ray tube and photograph the letters and figures produced on microfilm. The film can then be stored, viewed and reproduced in the usual way, albeit for a price. Such a system can handle 120,000 characters a second, generating 115,000 pages in an eight-hour working day. A 60 m (200 ft) reel of microfilm is equivalent to 4000 sheets of normal computer print-out on paper. It has been forecast that in ten years' time this business of computer output microfilming will account for nearly half the sales of equipment and film of the microfilm industry.

There are many other devices for communicating with the computer, ranging from typewriter-style keyboards, through plotters that can prepare graphs and diagrams, to television-like sets (graphic display units). At one time, these devices had to be within a few feet of the computer, but now computers are increasingly able to handle messages from remote terminals over standard communications lines, flashing back answers almost immediately. None of them, perhaps, has such fascination as the graphic display terminal which presents computer-generated data as lines of letters and figures (alphanumeric display), or as diagrams and drawings, or mixtures of both (Plate 4.2). These enable the airline clerk to check the up-to-the-minute state of booking on an aircraft; conduct the school pupil through a lesson; list the latest share prices in a stockbroker's office, and so on. Those that are equipped with a light pen enable the user to 'draw' on the surface of the tube: he has virtually direct rapport with the computer, which under his direction enlarges, diminishes, rotates, multiplies, stores, and edits his drawings (Plate 4.3).

Aspects of Software

Software represents the crucial frontier of the computer industry in the future. Each program is a uniquely precious piece of intellectual property, painstakingly constructed by specialists over many laborious hours. That means it is expensive, and incomprehensible to the uninitiated.

The sequence of instructions that tell the computer how to operate is known as the program (the spelling is widely recognized, with 'programme' being retained for the normal sense of the word). A computer can be programmed to solve any problem that can be defined in unambiguous, logical terms, but considerable human skill is needed first of all to dissect the problem and to serve it up to the computer in neat slices, with a tag on each telling the computer what to do with it. Hence the systems analyst first has to dissect the problem. That means not only defining what information is available and how it is to be obtained, but also what the program should accomplish—not always an easy task, especially when different people hold different views of what the computer is supposed to achieve. The analyst may also decide on the most feasible way of solving the problem.

The programmer next has to think through the analyst's statement of the problem and its solution. He expresses the solution in a series of logical steps, preparing it first in graphical form as 'flowcharts'. This is a vital stage in organizing the programmer's ideas, and is essential for explaining to other staff—who probably cannot understand programming language—what is happening. Finally, the programmer codes the solution from his flowcharts into programming language. In the early days of the computer, programmers communicated directly with the machine in the language they knew—machine code—but it was a very lengthy and expensive process. Computers cannot yet understand even stylized English, so the program that the programmer writes in a high-level language such as Fortran, Cobol or Algol has to be translated into the machine's code by a 'compiler' or assembly program. Fortran (Formula Translation) is written mostly in mathematical terms and some English and is used mostly for solving scientific and other calculations; Algol (Algorithmic Language) is somewhat similar, and is used mainly by scientists and engineers; and Cobol (Common Business-Oriented Language) is more appropriate for business needs. A typical program written for a small computer to carry out a simple business task such as calculating the payroll might run to 800 or 1000 instructions. These would be translated by the compiler into probably 3000 or more machine instructions. More ambitious programs run to tens of thousands of programming steps.

Then, using sample data, the programmer has to test the program on the computer, to see that it works, which rarely happens first time. As each fault occurs, he has to trace back through the program to the offending instruction and put it right. This is known as 'debugging', and can take a very long time on the more substantial programs. Another step which is usually necessary is for the programmer to document the program fully in such a way that others can understand the method of solution, so that they can be prepared to take over if he (or she) leaves in the middle of the job, as often happens in this mobile industry.

If programming has undoubtedly lagged behind hardware

development, a great deal more effort is now being put into programming languages and techniques. As computing becomes cheaper overall, and systems analysts and programmers more expensive, a greater proportion of programming will be done by the users—engineers, clerks, sales managers, etc.—instead of the professional programmer. Easier programming languages are the key. Though there are formidable problems in the way, the aim is to arrange for programming to be done in plain English or ordinary algebra. In time, almost anyone may be able to use a computer without having to learn a special set of techniques, a programming language, or even typing.

It is forecast that by about 1980 computers controlled by ordinary speech will be demonstrated, although the vocabulary and subjects may have to be restricted at first. A speech analyser would permit the computer to respond to such commands as 'Go', 'Start', and 'Stop'. At present, this form of communication is one of the fields of frontier research.

As electronic circuits become cheaper, mainly through the techniques of large-scale integration (see page 51), some software will be incorporated into the hardware. What are now instructions on punched paper tape will take the form of a plug-in unit of integrated circuits that perform the same function, so far as the computer is concerned. By 1980 it is forecast that the majority of software will be in this form.

Uses of the Computer

During the mid 1950s, at a time coinciding with the growth of first-generation machines, computers were generally used as super-calculators to solve difficult mathematical problems. In the office, they assumed the functions of girls with calculating machines. The first large-scale scientific use in the United States was in nuclear weapons research, a field that still commands some of the most formidable concentrations of computing power. Military needs also brought the first deployment of computers in monitoring aircraft movement, in the SAGE system in the United States, which employed some very large IBM computers (built in the heyday of the vacuum tube, and hence large and hungry for electricity) and also graphic display units.

The second-generation machines of the later 1950s and early 1960s were brought in on a widening variety of tasks, whose complexity and level of sophistication is enhanced as analysts and programmers become more confident in their skills. These tasks include data processing (EDP or electronic data processing), control systems and communications, and are edging into management jobs, areas that were formerly regarded as sacrosanct for human control.

The pattern will continue to change. According to a recent estimate of computer use made in the United States, some 70 per cent of the computer systems sold in 1968 were destined to carry out traditional accounting or record-keeping work; 20 per cent scientific and engineering calculations; and 10 per cent for production control, management information retrieval, and other kinds of 'operations' purposes. But in 1975 it is believed that this will have radically altered, with half the computer systems going into the 'operations' category, 30 per cent for scientific and engineering, and only 20 per cent for accounting, etc. But still, there will not be neat divisions between them and it will become more difficult to say with certainty into which class they fall.

The traditional accounting field, in which the great majority of companies with computers still engage their machines, displays the computer's great ability to store, sort, erase and re-record all kinds of information at high speed. The information may concern names, addresses, and customers' bank balances; medical case histories; part numbers in a factory store; daily production and cost figures in a workshop, and so forth. There are many routine tasks of this kind that the computer can do supremely well, alleviating the work of office staff: such tasks as calculating payroll, or controlling the level of a company's stocks of raw materials or parts. A computer enables a manufacturer of pop records, say, or the manager of a grocery chain, or of a chain of fashion shops, to regulate the daily orders at the factory, warehouse, or supplier by analysing the results of the previous day's sales. Only with the aid of the computer can he respond fast enough to keep up with the buyers' capricious tastes.

The computer is changing many other techniques of management, much as electronics, machine tools and mechanization have altered the tasks of production. Its most important feature lies in providing management with up-to-date, relevant information from all parts of the business. Armed with this information they ought to be able to steer the enterprise along with a much more responsive and a much defter touch than before.

One of the best things the computer has done for managers so far is to give them the chance of trying out different strategies on the computer without having to take the risks involved in conducting the exercise in the real, harsh world. In this kind of exercise they can investigate, say, the cost of putting up a new plant to manufacture a new product; can work out the cash they will have to spend and that they will recoup during various stages of the plant's life; the effect on the market and on the volume of sales of selling the product at different prices, all without venturing outside the haven of the computer's memory and logic. Once they have decided on a course of action, the computer—plus techniques of analysis and planning such as critical path analysis (CPA)—will help them to keep on course and remedy any deficiencies of scheduling. This is all the more valuable because these methods apply in any kind of management in which the problems can be quantified for the computer, whether the managers are in industrial companies, commercial firms, government or local authority departments, education and medical services, scientific research, or elsewhere.

Because the speed of the computer enables it to tackle many processes as they happen—so-called real-time operation—computer control of industrial plant becomes an attractive proposition where operations can be readily instrumented to provide the computer with its inputs of information, and equipped with actuators to control the process on command from the computer. The petroleum and chemical industries, whose flowline processes are among the most amenable for this sort of treatment, have been in the van in this application of automation.

The company that takes the next logical step of linking a data-processing computer to its process control computer (or computers) is well on the way to building up a computer hierarchy. The most advanced companies in several industrial countries are starting to do this now. In Britain within the next fifteen years all but the smallest firms in the process industries are likely to adopt some degree of computer control, with completely automated units handling continuous processes, and small computers optimizing the output from batch production. Some factories, and certainly refineries and chemical plants, will be entirely automatic before then, attended only by maintenance men for the great part of the time.

The scientist or engineer with a computer to hand is also in a powerful position both in his investigations of the natural world and when he comes to make useful things from its resources. Demand for computer time for scientific research in Britain—catching up after a later start than in the United States—is doubling every eighteen months. It used to be the physicists and mathematicians who set the pace; now increasing numbers of chemists, biologists, psychologists, economists and sociologists are turning to the computer for their work. Scientists always like to deduce simple mathematical laws about the behaviour of the world from their experiments, but have to introduce correction terms in order to simulate its complex behaviour at all faithfully. The value of the computer lies in the fact that it permits them to introduce virtually as many correction terms as they like, to construct huge arrays of equations in their mathematical models of the world (to simulate the weather pattern over northern Europe, for instance). These horrendous scientific expressions, which would be quite unmanageable by hand, bring the behaviour of the model nearer and nearer to reality. So subtle has this technique become that chemists can simulate chemical reactions in the computer, working purely from basic theory of atomic and molecular interactions; and physicists can glean accurate concepts of the structure of molecules from the patterns they obtain from X-ray pictures.

Engineers have long realized the value of the computer for solving the lengthy calculations involved in designing mechanical components, electrical machines, buildings, bridges—anything, in fact, whose shape and behaviour can be mathematically reduced to the computer's terms. Diagrams or drawings can be prepared by linking the computer to a graph plotter, graphic display, or other peripheral device. For example, from a set of equations that describe the structure and properties of a bridge, the computer can produce perspective drawings of the bridge seen from any angle, above and below. That is only one object: yet it may be a car, a ship, a drop of water falling through the air, or an 'electron map' of atoms—anything that is amenable to description as a mathematical model. One intriguing example is shown in Plate 4.4. Now that engineers have ascertained how to describe the shape and movements of a car in such terms, they can simulate a car crashing into a barrier and get the computer to draw out what would happen. Simulations of this sort are relatively cheap—and certainly less devastating—to carry out, and help to improve the design of cars, roads, crash barriers, and to strengthen accident prevention methods.

The value of a computer linked to a graphic display has been increased by making the computer able to respond to human action within the normal human response time—what is known as interactive use, since the computer and the human interact almost as though they were talking to each other, except that they are drawing instead of talking. Drawing with a light pen on the terminal screen, and pressing appropriate command buttons on the keyboard, a designer sketches in a car body shape on the screen, as seen in three-quarters front view. The computer then works out what the rear view would be like and displays that on the screen for the designer to see and amend. He can alter a radius here, make a structure thicker there, add a piece here, and take one out there. Once the design is finished to his liking, he can have the dimensions automatically stored by the computer for recalling whenever they are wanted. Or he may want it to calculate, say, the strength and stiffness of that particular structure from the stored data, and to compare the values obtained from this design with another one of slightly different form. When satisfied on that score, he can instruct the computer to prepare a punched paper tape that will guide a special machine tool to make that part.

In this interactive role the computer will have many jobs to do, ranging from design to education (where it can take students through their lessons without getting cross or flustered, and do so at each one's own pace). Considering the revolutionary nature of what is actually happening in machine terms, it is a reassuring, almost friendly, role.

But the effects of the computer will be felt far outside these perhaps esoteric fields. In medicine, computers are being used—especially in the United States—for keeping medical histories of hospital patients, and to remind nurses when to give the injections and doses of medicines that have been prescribed for patients. Some hospitals have used computers to keep watch on electro-cardiograph signals from patients with serious heart conditions who are maintained in intensive care units, and to analyse these signals to detect abnormalities which the doctor has programmed them to distinguish. In time, computers will be used to analyse clinical signs and symptoms, to assist the doctor's diagnosis, and, if the medical profession agrees, to take on the task of diagnosis itself.

Already many banks employ computers to process customers' transactions. The newest systems eliminate the bank clerk from routine transactions, as the customer inserts his bank card—like a credit card—into a cash dispenser unit, keys his identity number on the dispenser keyboard, and collects the money. The information is flashed to the computer, which up-dates his balance accordingly. At present most transactions involve pieces of paper at some stage—cheques, credit notes, and so on—but when all bank computers are interlinked it should be feasible to carry out transactions by simply transferring numbers from one computer memory to another. Each customer would have his plastics card with his code number on it, and would just present it when he wanted to 'pay' for something. The shop assistant would then insert the card into the shop's terminal, key in the amount, the date, etc., and that would be that. There would be no need to carry cash, as long as everyone accepted them. It would be a true 'cashless finance'.

Though most computers have to be coddled—guarded against shocks and wanton magnetic and electrical fields in carefully air-conditioned rooms—there are machines, pioneered by military services, that are robust enough to be taken into the field. Increasing numbers of aircraft, military and civil, are equipped with one or more computers to run their navigation, engineering and other systems; and as time goes by, it is highly probable that ships, trains, and even cars will have computers to improve the accuracy of their navigation and to enable them to travel more safely at higher speed. Computers have already turned up in 'unsuitable' surroundings such as steel mills and factories and on desk-tops in shops and offices, and, if cheap enough, may eventually be rented out to individuals for taking home, as terminals are now.

I said at the outset that the computer appears to have a certain affinity with the brain, although a comparison between them is rather far-fetched at present. The disparity is being reduced, however, both in terms of hardware and software, and the intelligent machine is taking shape. In the late 1950s and early 1960s there was a great enthusiasm, now rather diminished, for machine translation. This was one instance in which money and manpower were not enough to ensure a quick solution. In spite of large research grants that American government agencies started to give in the mid 1950s, chiefly for Russian and to a lesser extent for Chinese, the early work came to almost nothing because the message of the original generally got lost. Two examples are commonly quoted of how the computer can take the wrong meaning. The English proverb 'Out of sight, out of mind', translated by a computer into Russian, and then back into English as a check, came back as 'invisible idiot', while 'The spirit is willing but the flesh is weak' was perverted to 'The whisky is fine, but the steak's lousy'. The trouble is that real words have so many meanings. Yet at least early work on machine translation gave some impetus to the study of machine intelligence. So far, much of this work has been trying to demonstrate the basic feasibility of computers solving puzzles, playing draughts or noughts and crosses, answering questions posed in simple English, and recognizing objects in a picture; that is, constructing computable models of certain aspects of human behaviour. The last ten years have seen the creation of many interesting programs for conducting conversations, answering questions, and solving problems. In recent years a few laboratories have begun to build 'robots' that can see, move and solve problems by reasoning. At this stage they look nothing like the humanoid robots beloved of science-fiction writers, and are a long way from commercial application (although small numbers of fairly simple machines are performing straightforward jobs in factories, such as welding car bodies). But still, the foundations have been laid and the science of cybernetics (see page 152) has put the subject on a scientific footing. Perhaps in only another human generation computers will be able to match, simulate, or even improve upon some of Man's most striking intellectual features, and it will be feasible—perhaps attractive—to use these abilities in 'robots' of various kinds.

The Computer in Society

Even without 'intelligence', the computer is a formidable enough machine and it is making an extraordinary impact on society. That is to be expected, as it has broken the age-old human monopoly on the storage and use of information. Reactions to this fact vary. Immediate human problems are generated when a computer is brought into any kind of organization. There are bound to be tensions created by the installation of any new machine, but they are worse with a computer because it is replacing brainpower. This is a new experience for the staff, who will tend to rebel unless very carefully introduced to the proposition and kept fully involved in the project all the way along, and also reassured (if possible) that their jobs are not going to disappear.

One of the most potent sources of friction lies in the fact that the staff who minister to the computer are, in general, young. Even those who started with the computer industry in their thirties have barely reached their fifties. They feel more loyalty to the computer, which they know, than to the firm using it, which very likely they do not. Because they are in short supply, they can command high salaries, and constantly move from job to job to raise them higher. In the United States about twice as much is spent on staff as on hardware, and the trend in Britain is the same way. It is not uncommon for an organization that installs a computer to find itself with its third replacement team of programmers within two years.

The managers of today are most likely men who were educated and had their experience of technology before computers were invented—certainly before they came to be widely used. The chances of finding anyone outside the computer industry who has more than a vague, highly-coloured, biased and scanty knowledge of the computer and its uses are remote. It is a situation fraught with hazards for all concerned. The young men and women who understand the computer system tend to talk to each other—and, perhaps without thinking, to the managers—in their special jargon, so that the managers do not understand what is happening. This makes the younger ones impatient, while the older ones resent their precocious attitude and, fearing for their own jobs, try to hinder or even sabotage the computer people. At least one member of the company board will no doubt see the computer as an intruder, and rejoice when it fails. Some are not above giving it a push to help its downfall. Even so, they rely on the computer people to explain things to them, and have to trust them in the same way that they trust a lawyer or a doctor. Even quite lowly computer personnel can assume a frightening responsibility in handling data for their own or other companies, as they produce and use information which ordinary people cannot understand in its raw form by processes that are equally inscrutable.

Many fear redundancy when a computer appears. Reactions range from the furious—like the steel mill manager who threw all the computer's punched cards into the furnace—to the sullen and uncooperative. It is true that many information-processing jobs will be taken over by the computer, just as material-processing jobs have been taken over by the machines of production. However, this could be a blessing in disguise. The amount of information in existence, and hence available for processing, is rising at an astonishing rate, and if we do not all want to end up working in offices handling it, we shall have to make the most of the computer, and other devices that handle information. Already some 13 per cent of the working population in Britain work in offices. The introduction of computers on a large scale will allow the control of industrial, commercial and government records to improve without further massive increases in office staff. It is predicted that by 1974 computers will have taken over 9 per cent of office jobs in Britain, compared with only 0·75 per cent in 1964. True, those who would otherwise have worked in offices will have to find other jobs, but, all in all, it is forecast that even by 1975 the number of additional office jobs created will still outpace those taken over by the computer. Even staff reductions, if carefully planned, can be accomplished largely through cutting recruitment and not replacing those who leave, and involve little real hardship. Besides, extra staff will be needed to gather and prepare information (e.g. punching cards), to run the computer itself, to receive and distribute the processed information and to take over again if the computer system goes wrong. Often there is more work to do in correcting the results of errors in the computer system than ever there was before.

Sometimes mistakes are hilarious—like one bank's computer which said a man had an overdraft of £8 million. There are numerous stories of the computer sending out demands for the payment of a bill for £0·00, and then approving the payment of a cheque for that amount. Cumulatively, other instances have an undoubtedly sinister ring. In one case a research chemist with a large international company fed in a scientific program to the computer and was staggered to find that the computer produced not the results of his calculations but a flood of confidential information from the company's personnel files which were kept on the same computer. Perhaps only a trivial example, this does show how the wrong information from a computer 'data bank' may be given inadvertently to an unauthorized user. With the accretion of larger and larger 'banks' maintained by central and local government there is a growing danger of this happening not only within the government machine, but also at the behest of others who want information for the wrong reasons. It is quite feasible that data banks could be tapped and the information disclosed for blackmail. While it should be feasible to introduce checks—a sort of lock and key—into the program so that only users with the right code could gain access to certain types of confidential information, the anxieties about improper use of personal data still remain.

Thorough preparation for a computer's installation is essential. The apparently casual fashion in which many companies launch on computer projects is almost unbelievable, although sums of tens or even of hundreds of thousands of pounds are involved. Companies have been known to order a computer and then leave it unconnected in a prominent place so as to impress visitors. Even those that try to make adequate preparations but fail, probably through inexperience, may well be using the computer inefficiently five years after installation, and will regret it for much longer. A survey carried out in Britain in 1968 disclosed that nearly half the users felt they were not getting a proper return from the money they had invested.

The trouble is that the organization must be tidy before the computer can be applied successfully to its affairs. Few are. The most difficult task for the systems analyst, programmers and company personnel before the computer arrives is to unravel the organization's tangled web of operations and communication and then to lay it out in logical sequences of command and action for the computer. The machine will have no experience of the business and no sense of the ridiculous or the obvious. There are likely to be many things that people take for granted and that no one remembers to put in the program but are nevertheless vital for its success. Everyone has to think hard about what actually does happen in the organization, as opposed to what is supposed to happen, and what the aims of the work are. Often these aims have never been defined.

Clerks were the first ones to feel the effects of the computer revolution, as various aspects of accounting, sorting, filing and transmitting information were taken over. Now it is the turn of managers and administrators. They will claim vehemently that their work involves so much human intuition, flair, hunch and the rest of it that it can never be successfully computerized, but their protestations will almost inevitably be in vain. As computer staff become more skilful, managers should feel more confident in giving the machine more difficult, diffuse problems to solve. In future, many of the factors that are now said to need human understanding will be reduced to mathematics, like the others.

The computer revolution is well under way. The computer will become the most basic tool of technology in the future, fundamental to every kind of human endeavour. It will eventually transform the nature of work, removing much of the mental effort involved, and people's attitudes to their work. It will permit (indeed, encourage) large organizations to flourish by providing central offices with the timely, critical information they need to exercise control. It should enable administrators to take much more account of the social effects of technology before they are precipitated by Man's actions, and mitigate them. Even in the United States, the experience of using computers is a new one, and so we cannot really visualize what they can do. This is only natural with a new technological device of such power. In time individual computers, or terminals, will be regarded as essential for home, business, profession and school, and the ability to use them well will be as highly thought of as the ability to drive a car is today.

In the short period that Man has had the computer, he has used it to enable him to mount some of his biggest projects—to get to the moon; to run international businesses; to control weapons systems around the world. It has been a short period, too. To go back to our imagined year of human development (see page 40), the computer has existed for about seven minutes. In the next seven the effects of the computer will really permeate through society. After that it is a matter of choice. We have nothing to fear, though, and a great deal to gain, if we learn to understand it and handle it properly.

5 · Mass Production

Productivity and Possessions

We are in an age of mass production and mass consumption. The paraphernalia of modern living—cars, milk bottles, electric razors, plastic buckets, washing machines, synthetic fibres, gramophones, light bulbs, clothes, and the rest—pour forth in unending profusion from the factories of the world, churned out in tens of thousands, and ultimately millions, for Man's comfort and convenience (Plates 5.1, 5.2). In order to maintain this flow, technology is ripping out the riches of the environment and turning them into adjuncts of technological life that are soon discarded. Markets for finished goods, sources of raw materials, and manufacturing operations are exploited on an international scale. Thus a radio set made in Britain may incorporate Zambian copper, Australian silver, aluminium smelted in Canada from Jamaican bauxite, ceramics prepared from English clay and Norwegian feldspar, Indian mica, housed in a plastics case made from oil from the Middle East, with a handle of stainless steel based on iron smelted in England from mixed Swedish, Brazilian and West African ores, incorporating Canadian nickel and Turkish chromium.

Other products are just as cosmopolitan. Take cars. Often the parts are made in one country and assembled in another. This practice is so common that the question of whether a car is British, German or French is often in doubt. Technology transcends national boundaries. Mathematics is a universal language, and, provided everyone is working to common standards, an engineering drawing or a punched paper tape can provide the outline and details for a workable machine anywhere in the world. Metric systems of mensuration form the common basis for technology in most industrial countries, with the notable exception of the United States, where a change is to be expected before too long.

Although we now take it for granted, mass production is a comparatively recent part of technology. It began, as we might suspect, in the United States in the 1920s and spread to Western Europe, the Soviet Union, Japan and China. The more well-to-do people bought their first telephones, radios and cars; they discovered a host of goods that had either not existed before or had previously been too expensive, and eagerly snapped them up. Firms grew rapidly and new ones were established. Industry conquered the unexploited market, while technology gathered the almost pristine resources of the globe and started the first wide-scale pollution of air and water. By the 1950s, with the intervention of war and the diversion of some production into arms and military supplies, some two-thirds of the American population had bought all that they could reasonably want or expect. This set a limit to the amount that could be sold, even if it could be made. It marked the beginning of the affluent society in the industrialized countries—the others quite soon reached the same position as the United States—in which consumption of goods had to be kept at an artificially high level in relation to people's needs. Advertising and other means of persuasion had to make people throw away possessions more often, and in greater quantity, to make way for new and often more costly ones. The goods themselves were often made of thinner, lighter materials, welded or glued together rather than bolted or screwed, as a result of industry's continual search for ways of cutting costs of manufacturing and materials. Hence goods started to wear out more quickly, and the manufacturers gained their replacement markets very easily.

The adoption of simple mechanization and the spread of mass production techniques helped boost productivity and output. In the United States, for instance, productivity rose by about 300 per cent between 1919 and 1963 while employment increased only by 170 per cent. To handle the growing amount of administration involved, more workers moved into the office. For every 100 production workers in American industry in 1919 there were 19 or 20 non-production staff; but by 1963 there were 26. The main part of the increase occurred in the last decade of that period.

The Heart of Technology

The basis of all manufacturing industry, whether it deals in mass production or not, and also the heart of technology, comprises methods of shaping materials and joining them together. Man has always been a tool-maker ever since the flint-chipping days, and few factors can have played a bigger part in determining the rate at which he has exploited his environment than his ability to make and use tools. Shaping can be done by cutting, by eroding, by electrical action, or by moulding, usually under heat and pressure. The machine tools that carry out these methods of shaping have changed little in basic design for twenty-five or sometimes a hundred years. The main advances have been won by raising their power so that they can hack away metal faster, aided by cutting tools tipped with tungsten carbide, a material which in this context has been one of the main aids to productivity in the last twenty-five years. To stand up to it, the machines have become heavier, stronger, more rigid and capable of running at a greater variety of speeds. Setting-up time has been reduced with the aid of pre-set tooling while loading has also been speeded up.

But the new materials and demands for greater efficiency have started to upset these traditional techniques. If a certain amount of adaptation to old shaping methods is possible in going from old materials to the new, in many cases something more radical has to be devised. Often the material—a ceramic, say, or the nickel-chromium steels—will be too hard, too brittle or too reactive for conventional methods. Even if one of these old methods is modified, it is generally too slow, too wasteful of materials and too dependent on a craftsman's skill and patience for today's mass-production era. As materials change, so technology has to change with them.

There are many possible processes available. The materials can be moulded into shape from a powder where appropriate—like iron powder, or silicon carbide—or they can be shaped by a variety of new methods that employ not the direct action of a cutting tool but the influence of electron beams, magnetic and electrical fields, and electrical corrosion or plating.

A case in point is electrochemical machining. This technique is virtually the opposite of electroplating. Instead of material being electrically deposited on a metal, the metal itself is eaten away through the action of an electric current. The workpiece and the conducting, shaped tool are connected to opposite poles of an electrical supply (the work forming the anode and the tool the cathode) and a heavy current is passed between them. An electrically conducting solution, the electrolyte, is pumped at high pressure through the gap between them and carries away metal eroded by the current (as shown in Diagram 5.1). This allows extremely precise, complex holes to be cut in very hard, brittle

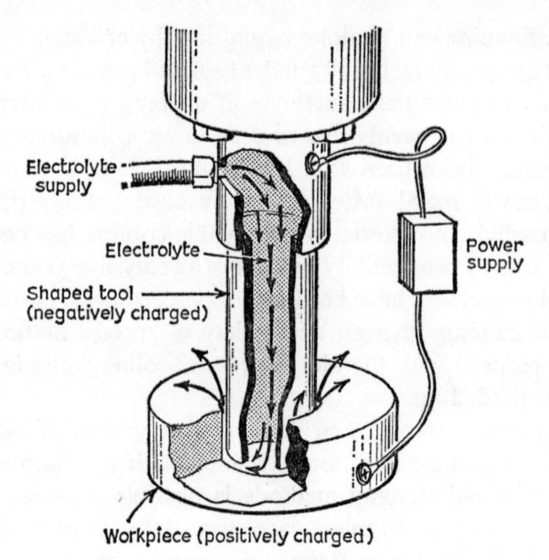

Diagram 5.1 Electrochemical machining set up

materials, the hole being the same shape as the cross-section of the end of the tool.

Ultrasonics provides another useful tool. Sound waves above the audible (above about 16,000 Hz) can be used to cut brittle materials, such as glass, and to weld metals and plastics. They can detect flaws inside the body of a metal sheet or casting, since they bounce off the flaw and give an echo that is detected and recorded. Frequencies from about 0·25 to 10 MHz are commonly used in flaw detection. In a steelworks billets can be automatically scanned by ultrasonic probes as they come up to the rolling stands to be rolled into plate, girders, etc. It saves a great deal of time and money to be able to spot and discard faulty billets before they are rolled into defective products. The illustration (Plate 5.3) shows an ultrasonic detection device being used to check the quality of pipeline welds in the field.

Another cutting and welding 'tool' is the plasma flame. Though the normal oxy-acetylene flame is hot enough to be widely employed to cut steel plate, the plasma flame is hotter—at up to 15,000°C it is hotter than the surface of the sun and far outstrips the oxy-acetylene flame's 3400°C. This is high enough to cut through specially resistant materials like asbestos cement, molybdenum, tungsten and ceramics.

The trouble with these methods of shaping is that they waste such a large proportion of the material as scrap—machine tools commonly turn out 50 per cent of the original workpiece in this form. Though the scrap can be returned, and indeed scrap is a major source of raw material for the steel industry, the machining process is obviously wasteful. Alternative methods that shape the material without cutting are preferable on this count. Often conventional techniques can be suitably adapted without too much trouble, such as pressing the material between a shaped die and punch or squeezing it like toothpaste through a shaped die to make a rod or bar of metal or plastic. Precision casting, where the metal is cast in a mould that accurately reproduces the final shape so as to eliminate further machining, is another successful adaptation. Forging—the way in which a blacksmith forges metal on an anvil—has its modern variant in high-energy rate forming, which was pioneered in the United States in the mid 1950s. As the name implies, it shapes metals very quickly, at far higher speeds than those encountered in ordinary forging. They, like the other new methods, have their strong points but they do need special skills and are found comparatively rarely in production processes, and very rarely in mass production.

Jointing methods are also changing under the impact of technological progress. Perhaps the oldest method that Man learned was tying a stone on a wooden haft with a string of grass to make a primitive hammer. From then on, jointing techniques kept pace with advances in materials and in shaping techniques. The production and use of electricity on a wide scale through the nineteenth century prompted the development of electric welding. In the 1930s, when aluminium alloys were widely adopted for aircraft structures, inert-gas arc welding processes had to be devised for joining them, the gas (e.g. argon) protecting the easily oxidized aluminium from the air. Structural failures of welded ships built during the Second World War stimulated a great deal of research on the brittle fracture of steel and other metals. This in turn led to the development of high-strength materials, an example of how advance in one field of technology stimulates that in another.

As a result of this continuing development, there are now about forty different methods used commercially for joining metals ranging from arc welding, to solid-phase welding, to adhesive bonding where joints are held by molecular forces between the surfaces and the adhesive. Then there are all the other methods which involve a mechanical means of attachment, such as rivets, bolts, screws; and hosts of different kinds of fasteners, among them zips, buttons, adhesives tapes, and so on. The field is extremely complicated, and as each new material comes along it demands an adaptation of, or improvement on, existing joining methods, complicating it still more.

Inevitably, new processes such as these, while necessary for dealing with the special materials of modern technology, leave much of it untouched. Changes come slowly to mass production, which can only accept innovation in small doses and where mistakes with untried methods can be extremely expensive. While iron and steel remain the most important materials for technology, machine tools using conventional processes will remain the most important devices for shaping them.

That is not to say that these processes cannot be improved. Far from it. Traditionally, one machine tool has performed a limited number of tasks—such as drilling, or milling only—and a lot of time is wasted while the work is transferred manually from one machine to the next and the tools set on each. The tendency now is to equip one machine with a set of pre-set tools on one head, each of which can be brought into play by simply revolving the head, so that a series of operations are carried out on the one machine. This kind of machine—a 'machining centre'—is able also to turn the work so that a cutter can get at all its faces, apart from that on the machine's table. Such a machine carries out a sequence of operations that would otherwise involve a succession of machines of more limited scope.

Such multi-spindle work centres are particularly suitable for automatic operation, and this is conveniently carried out under numerical control (N/C for short). Essentially, this implies that the information needed to make the raw metal into a finished component is put on to punched paper (sometimes magnetic) tape by a skilled programmer, and the machine tool shapes the piece under the control of that tape and without intervention, or with only minimum intervention, by the human operator.

Though these N/C tools cost more than conventional machine-tools, they do not need the special jigs and fixtures which normally have to be designed and made to hold the workpiece and guide the tools on to it before starting production with conventional machines. All that is required is a simple clamping device to hold the work. Numerical control also raises productivity because it increases the time the machine is actually cutting metal, typically up to 40 per cent as opposed to only 20–25 per cent for manually operated machines. Besides that, N/C often brings improved quality and more consistent work, and makes it possible to machine complex shapes that would be very difficult to do with manual control. All this leads to faster production, and cuts the time taken to make components by a significant amount. One N/C machine can often produce as much as two or three manually controlled machines in a given time. There may be other incidental benefits, such as reducing scrap.

The first N/C machines were introduced in the United States about 1953, and by 1967 their total had risen to around 10,000, compared with about 2300 in Britain.

The logical extension of this sort of operation is to have a line of machines, each operating automatically under the control of its own program of instructions, shaping parts in succession, so that a blank is fed in at one end, transferred automatically from one machine to the next for the different shaping operations, and finally emerges as a finished part at the other, untouched by human hand. All that the human operator would do is to prepare metal blanks and fix them to pallets which would then travel through the system with the metal pieces on them. The whole system would be under the control of a computer.

Looking a little further into the future, it is not hard to envisage the factory that is practically automatic in all its operations. It would be run by computer—probably a computer hierarchy—and would need people only to supply it with raw materials (in such forms as sheets and bars of metal), to take the finished parts out, and to maintain the production machines and the computers. That would be automation in its most fully developed form.

The Shape of Automation

This vision of technology is not yet realized, however. Before we go on to consider its social effects, let us look in a little more detail at what automation is. This is difficult because there is no satisfactory definition. In the United States the term tends to be accepted for any piece of machinery or any new way of doing a job that displaces labour, whereas in Britain it would often be called mechanization. Hence the confusion over its effects.

Full automation implies a total absence of human beings—intelligent machines running their own affairs—and although this is a possibility for the late 1970s or early 1980s, a long series of gentle steps lies between current technology and full automation.

Various techniques grouped together under the general heading of 'low-cost automation' can increase output from machine tools, say, by the simple expedient of fixing perhaps digital indicators and stops to the machine so that the operator does not need to spend so long on setting it up, checking the position, and so forth. He can read the position of the tool in unequivocal figures on a large dial and does not have to measure dimensions with calipers as he goes along. A more advanced and more expensive system reduces the need for sustained human attention considerably, and the machine (e.g. an automatic lathe) makes components under the guidance of a control unit with a pre-programmed sequence of instructions.

Very often parts that are being produced have to be transferred from one machine to another for various processes to be carried out on them, and here automation has much scope. Take the case of a casting for a car engine block, which starts life in the foundry and is converted into the finished component through a variety of milling, drilling and broaching operations. We could evolve an automated line for this in steps, starting with a series of machines, each carrying out one operation on the casting, each tended by an operator who loads and unloads it and brings the tool down on the casting by pulling a lever. Then, by arranging to have the movement of the tool controlled by stops, the operator only has to push buttons and load and unload the machine. Next, the introduction of multiple cutters or drills allows each machine to drill many holes—or make several cuts—at once, increasing the production rate considerably. A roller-conveyor installed between the machines reduces the time taken for loading and unloading. The big step comes when all the push-buttons for all the machines are grouped in one control box, so that they can be controlled by one man, and electrical interlocks prevent one machine starting until all are ready to start and there is a casting in position at the beginning of the lines. Finally, by linking custom-designed machines (to replace the standard processing machines) with a conveyor which is timed and set so that the casting is automatically transferred between, and presented to, the machines at the right time, we have created a transfer machine.

This could be handling castings; it might be filling and capping milk bottles or cans of peas. It might be making complete cars. In fact, it might be performing any one of a multifarious range of activities of mass production. The machines might be arranged in a circle or a square, or simply in a line, so that the work either goes from one end right on to the other, or comes back again to the starting point. The circular layout is frequently adopted where the work is mounted on a platen for its journey through the line, the platen being returned at the end and used again for a fresh work-piece. In its condensed form, the circular layout may be simply a rotating wheel, or table, with machines mounted on a central pillar or grouped around the edge. There are many variations on the type of machine required, the manner in which they are arranged, the amount of manual handling needed, and the points at which the quality of the product is checked, depending on the job.

The basic production process consists of making parts, assembling them, and then finishing and testing the assembly. Mass-production companies often find that more man-hours are spent in assembly and testing of the completed assemblies than in making the necessary parts. In fact, assembly may account for as much as 60 per cent of the manufacturing costs of many consumer goods. Again, because metal-working has been made so efficient, it is difficult for an efficient company to lop 10 per cent off its metal-forming or metal-cutting costs, and so it has to turn to other operations such as assembly. There are sound reasons, then, for mechanizing this operation, which in Britain alone occupies one-third of the total of 4 million people employed in the engineering industry.

Much of it could probably be done by machine but the human being is so good at the job that at present it is not feasible to make an assembly machine with the human operator's skill and adaptability. The process, and the part, therefore have to be designed in a way that existing machines can cope with. For automatic assembly to work smoothly, the automatic feeder that supplies parts from a store must be absolutely foolproof: the part must be brought in the right way up at the right place at the right instant. Any failure can jam the whole production line. In this context, 'part' is interpreted liberally; it may be literally a part, such as a bolt, or it may be a certain weight of sugar to put into a bag, or inert gas for filling an electric light bulb, or even a quantity of molten solder for securing electrical connections. Next the part has to be placed at the proper point in the assembly, and finally secured in place (or whatever action is appropriate to 'secure' the sugar in its bag, for instance, or to seal the gas in its light bulb). The machines that are needed to simulate even apparently simple human actions like picking up a bolt and inserting it into an assembly are very complex, and apart from the problems of designing and making the machines themselves are the subsidiary ones of training operators to use them (some trained people are still needed), and preventing them being jammed by faulty components, odd bits of rag, or even by oil gumming them up if the parts themselves are light.

Nevertheless, the shortage of labour in industrial countries and the rapidly increasing wages bill are forcing manufacturers to look for as many ways as possible of replacing personnel with machines at all stages of the production process—in production itself, in assembly, and in finishing and testing, besides ancillary services such as storage and materials handling. Until recently, the major effort in mechanization has been devoted to mechanizing mass production—which implies a production rate of at least 1 million items a year, continued for several years. Such a scale of operation affords the chance and the justification for building special-purpose machines designed and built to do only that particular job. However, even in the United States, the home of mass production, more products are made in medium- to small-scale production than by mass production, and now attention is turning to the mechanization of these less substantial operations.

To make the fullest use of automation there must be some form of remote control of a machine and remote sensing of its output. Early on in the history of mechanization, what are now called cybernetic techniques came to be used. In a cybernetic system there is some method of looking at the output of the system, comparing it with a desired norm, and then using the difference to actuate a control device that tries to bring the system back to the desired state. A thermostat is one example. Another is a steam engine whose speed is controlled by a governor. There are many kinds in operation throughout industry.

Take the case of an electronic control system, which we can compare with a human nerve system. Transducers (sensing and measurement devices) approximately correspond to the five human senses; a computer in the most advanced systems, and some other electronic control device in less advanced ones, corresponds to the brain; actuators correspond to muscles; and the various units are connected by wires, analogous to nerve fibres. Transducers convert such quantities as pressure, temperature and liquid flow into electrical signals which can then be processed by the circuitry, while at the other end of the chain of command the actuators convert them back again into some other action—usually a mechanical one, such as opening or closing a valve, or starting an electric motor. There are various kinds of cybernetic systems: wholly electrical; pneumatic; or hydraulic; or combinations of these. In the last few years there has been a resurgence of interest in 'fluidics', the pneumatic equivalent of electronics, which performs the same sort of control functions, using air or other fluid instead of electric current. Because of their robustness, fluidic devices are attractive for military purposes—in missile guidance systems, for instance—but they have other properties, including resistance to high temperature and radiation, that make them suitable for certain industrial applications like controlling machine tools and process machinery. Each type of system has its advantages for a given kind of job, and they may often be combined. Often there will be two sensors for one task. Let us consider the case of an automatic machine filling and capping bottles. One inductive probe would be mounted above the bottles on a conveyor, while a photocell system would shine light across the bottles. As the bottles move along the line, one of them at some instant crosses the light beam, and when it does, the cap should be immediately below the probe, and so both sensors should give a signal. If they do not, something is wrong—the bottle may be without a cap, say—and corrective action can be taken. The job could be done by a person, but the machine is much better.

Other tasks could not be handled manually. Many of the processes—especially chemical—that convert raw commodities into useful materials happen so fast and are run so close to instability (or runaway) that a man is too slow to control them. Some form of instrument control, with rapid response, is necessary to keep them running safely. These plants are comparatively easy to convert to computer control, as the control system is all set up ready, and what remains to be done is to put a computer instead of a person at the summit of the hierarchy. The oil and chemical industries are among the leaders in this form of automation (see page 94). Other processes are less easy to convert. Yet in order to reap the fullest benefits from automation, the managers of any sort of technological activity will increasingly have to rely on computer control.

Mass Production—Value and Obligations

The reason that the manufacturer likes mass production is that it lowers his costs for making a given article, and that is a big factor in all production technology. While a machine is running it is earning its keep: while it is standing idle, it is losing him money. The longer a production run of a given article, the less time is wasted changing the machines and making the different jigs, tools, dies, and so on that are required to produce new shapes. Also, the fewer are the different items he has to store, which can bring substantial savings of its own in terms of floor space, storage racks, handling equipment and staff. A constantly-changing type of article—and there are some, like cars, that change often—can make great demands on storage space for keeping a wide variety of replacement parts. The more articles of the same type that the manufacturer can produce in a given time, and the longer he can keep that article in production without modification, the better. The more articles he can sell in a given time, the greater his turnover and the greater his profit. Mass production means that the articles can be sold for the same price, with greater profit for the firm on each, or they can be sold at a lower price with the object of stimulating sales. The size of the market depends on price. A cheap colour television set, for example, would be bought by far more people than buy or rent sets today.

The same economies apply to plants that produce chemicals, such as detergents, drugs, plastics, and so on (see page 73). Capital costs for many process plants are found to increase less rapidly than their capacity—in fact on average the cost is proportional to the capacity raised to the one-sixth power. Hence the rule of thumb is that a 200,000 ton per year plant can be built for only about 50–60 per cent more capital than would be required for a plant producing 100,000 tons a year. Besides that, it takes merely a few more men to run and maintain the larger plant than the smaller. Also, the threat of technical obsolescence means that the plant may have to be written off after ten years. Hence depreciation is heavy, and the higher its rate of production, the more profit it can earn to offset that before it has to be scrapped. For all these reasons, production on a massive scale is well worth while.

However, when production is on a massive scale, so are the results of any miscalculations. Every month's delay in bringing a large new chemical plant on stream could lose as much as £500,000 in profit. Previously, when plants were smaller and simpler, a company would fix the expected selling price of the product according to what it thought could be sold, and what its competitors were doing. But now one plant may produce so much that its output actually alters the price at which the product is sold all over the world. When really large plants are brought into production, they may be too large for the existing market, and for a while there is too much product and so prices slump and all the producers lose. In time, the growth of the market allows it to catch up with the capacity again. So that the large scale in production works both ways.

Mass production is necessary to cater for mass needs; but it pleases some and exasperates others. For a start, it implies standardization of products. Machines cannot be continually altered to make variegated products to suit the whim of individual customers when the factory is geared to the production of millions of identical items. There is always a conflict between the regimentation of the production line and the customer's hankering after more variety and a wider choice.

It makes stringent demands on the technology of production in addition. One thing is vital: that all the parts should fit together, with no misfits and no gaps. All parts of one type must be identical within certain prescribed limits. At one time machines and products were all made by skilled craftsmen, one at a time, each part being matched to the others by painstaking effort. Batch production dates back to the end of the eighteenth century, when the American Eli Whitney managed to get this interchangeability of parts for firearms, but true mass production only became feasible when another American, Henry Ford, realized that accuracy was all important and applied this realization to producing large volumes of cars. He told his engineers to learn to measure to a millionth of an inch; accurate enough to mass-produce efficiently. Now all the millions of parts from the production line must be produced to an accurate standard, if not that accurate, then often to a matter of thousandths of an inch. This involves precise measurement and close inspection, reliable raw materials and accurate setting up of the machines making the parts (Plate 5.4).

It also demands high speed so that the plant can turn out as many items as possible in a given time. This places a heavy burden on the production machines and on the organization of the factory and its work. Techniques of production engineering have to be employed to wring the last drop of output from men and machines. Machines need to be reliable so that they can run for long periods without breakdown, and factors of reliability, maintenance and environmental engineering (investigating the interaction between the machine and its surroundings) assume a great importance.

Mistakes in the design and manufacture of the product itself are more costly and affect more people. As machines tend to become 'systems' with large numbers of interacting components, reliability becomes more vital and the consequences of failure more drastic. Today's products are more complicated than ever before. A modern aircraft engine, for instance, may contain 30,000 individual components of over 4000 different types. A modern car contains something like 15,000 parts. An intercontinental ballistic missile (although not typical of a mass-produced product) may contain 300,000 different pieces, any one of which can jeopardize its success. This complication multiplies the difficulties of management. The company designing and making the main product may well depend on materials, components and services from anything between 200 and 4000 suppliers and vendors in its larger projects.

The larger the firm, and the more freely its products flow across the world, it has to watch its step that much more carefully. A mistake can affect hundreds of thousands of people who have bought its products. A fault in only 1 per cent of the tinplate supply from a steel mill, for example, may result in 50 million faulty cans a year. A large car manufacturer may have to recall a million cars to correct a small design fault. Mechanical failures are predominantly due to fatigue (weakening from repeated load cycling) and, to a lesser degree, creep (deformation under prolonged load). Often failures are brought about because the design is inadequate for the purpose, because the designer has not been told enough about the conditions and surroundings in which the device will have to work, or because the validity of the design has been inadequately tested.

Mass production also demands more machinery. Companies are steadily installing more equipment for each man or woman they employ, and as a result their businesses are becoming more capital-intensive. The engineering masters of the nineteenth century could get by with around £10 investment in plant and machinery per man; now the total would be more like £2000, and much higher in some cases. In chemicals, for instance, the figure for a large British group was no less than £9000 per employee in 1969. The higher the figure, the more advanced the technology is likely to be, the greater the degree of automation, and the more extensive the replacement of manual effort by machinery.

Human Cost of Automation

Mass production, aided by automation, delivers the myriad products of technological life in far greater profusion and at lower cost than would ever have been possible by the manual methods of the pre-1920s. Yet it has its human toll. No one has yet been able to measure it exactly. Different experts have different and often wildly divergent opinions, even on the most important, and apparently simplest question: how many jobs are being destroyed by automation? The American manufacturers of automation equipment have estimated that automation in all its forms is depriving their country of 40,000 to 50,000 jobs a week, or two million or more in a year. In the US steel industry alone, limited automation is believed to have eliminated 80,000 jobs out of a total of 600,000 between 1953 and 1966, with at least another 100,000 to go in the next decade. The automation industry itself is growing rapidly, and in the United States, its sales are expected to exceed $30,000 million annually by 1974, compared with around $22,000 million at present. The sales cover such items as electrical switches and meters, industrial instruments, domestic and commercial building controls (for heating and ventilating, lighting, etc.) and aerospace and defence control systems. Within the total, the provision of computer terminals, computer services, data communication equipment, and test and measuring instruments are among the fastest-growing sectors—some of them expanding at 40 per cent a year and thus reflecting the growing importance of information to technology. These growth rates will be boosted by the adoption of integrated circuits which allow such products to be smaller, lighter, longer lasting.

Automated—or simply mechanized—systems do away with jobs on an increasing scale. So that operators vanish from automatic lifts and automatic telephone switchboards. Warehouses and stores equipped with the latest materials-handling aids, such as forklift trucks, need fewer men to fetch and carry. The self-service trend has reduced the need for assistants in supermarkets (none, of course, are required for vending machines); while self-service laundries and 'easy-care' fabrics have bitten into the traditional laundry trade. Fewer post office workers are required to install telecommunications equipment, and to sort mail, as machines take over certain aspects of these tasks. And so on.

Automation has displaced many semi-skilled people in the mass-production industries. In the last decade assembly lines have been automated and the total labour force has dropped a long way while output has continued to grow. Machines are steadily taking over more of the production process. In fact, the verb 'to manufacture', with its literal meaning of making something by hand, is becoming increasingly inappropriate. The production line is an international facet of technology, whose appearance varies little across the world.

The spur to mechanization is naturally the desire of all industries and other organizations to reduce costs, and at a time when inflation and strong labour demands are pushing up wages and salaries rapidly, that spur is felt even more keenly. Typically about 70 per cent of the cost of making a motor car, a radio, a pair of shoes, is taken for the wages of those who have helped to make it, and a machine is generally cheaper in the long run than the man (or usually men) it replaces, and is better worked harder. Besides, the machine is never late for work—although it may fall ill—cannot answer back, and does not leave to go to another firm for more money.

All this means individual hardship, and in some cases individual hardship many times over, but we must be careful not to read too much into the figures. A skilled man will seldom be declared redundant. In most industrial countries there are normally more vacancies than there are right people to fill them, and skilled men especially can almost always land on their feet. In any case, outright firing because of technological change is not all that common. Usually displaced employees can be kept on in similar jobs (especially if the installation of a machine raises the firm's productivity), or redeployed within the organization. An organization that is aware of technological change will be able to plan well ahead of a new machine's arrival, and to run down its labour force beforehand deliberately by not replacing those who leave in the normal course of events. It seems likely that economic growth and the rise of new industries (even of the automation industries themselves) create more opportunities than there are jobs destroyed. At any rate, taking the industrial world as a whole, unemployment rates are generally under about 5 per cent of the working population, and consistent with what the economists call 'full employment', despite the onslaught of automation. There is no evidence from the experience of the last 150 years that, in the long run, increasing mechanization brings anything but shorter hours of work and higher living standards.

In addition, automation is spreading much more slowly than the more ardent prophecies would have us believe, notably because the necessary machines are so expensive and there is a shortage of skilled people to make, operate and maintain them. In the long run too, only a proportion of jobs would be mechanized. It would be difficult—maybe impossible—to replace human beings with machines in many kinds of low-skilled task on building sites, in shops, offices, factories, homes, etc., and in highly skilled functions like conducting international diplomacy (as much in business as in government) and such tasks as nursing or teaching.

More often the hardship comes in adapting to the change that technology brings: an old hand has to learn new tricks. The growing use of increasingly complex machines leads to demands for more skilled people to make, operate, maintain, repair them. Operators, maintenance staff, repairmen, supervisors, salesmen and managers all have to be trained to a higher level in order to appreciate what is going on inside the machines and understand how they are used, and how to use them. In many areas of production, a skilled operator can keep a machine running under normal circumstances, though a 'trouble-shooter'—a maintenance man with an all-round knowledge of electrical and mechanical engineering—will be needed to take over when the operator gets out of his depth. Frequently the major obstacle to introducing and successfully running a new machine is the operator's fear of not being able to handle it, or of losing his job if it works properly. Besides needing to be trained to handle it competently, he needs to have his fears dispelled (if there is no cause for them) or treated honestly (if there is).

As a consequence of technological change, a man will have to be retrained to stay with the same organization; or if the worst comes to the worst and he is finally displaced, retrained to take up some other form of work elsewhere if his old skills are outworn. There will be hardship because people—the older ones and the semi-skilled are likely to be hardest hit—have to move from the location of old industries to new, even where they can learn the new skills and are prepared to make the effort necessary to get to the new place of work. It is a serious personal and social loss, but with greater official awareness of the problem and the willingness to help displaced employees, and a more receptive attitude to learning on the part of the employees themselves, the transition can be made less painful.

The Giant Enterprise

In this mass-production, mass-consumption age the individual can accomplish little on his own. Power resides in the large organization—government, company or international agency—in which individuals can collectively exert enough influence to affect the environment markedly, altering it to conform with their way of life, and converting the world's natural resources into useful products on the massive scale required. Such organizations must rely on computers and rapid communications to keep them supplied with essential up-to-the-minute information if they are not to be struck with a creeping paralysis bred of sheer size. Once they have these facilities, however, size is no hindrance. Indeed, it is an asset, and one that any enterprise needs if it is to operate on the global scale commensurate with modern technology.

A colossus such as General Motors—employing 600,000, and doing business to the tune of £10,000 million a year, about equal to the gross national product of the Netherlands, and larger than Switzerland's—bestrides the international scene, finding its raw materials and selling its products all over the world. There are many giants of the kind. They compete with other giants for raw materials, markets and manpower across the frontiers artificially imposed by governments, which prove increasingly irksome (see page 337). Only the large firm can indulge in the production of millions of cars, or millions of tons of chemicals, or whatever the product is, thus reaping the economies of scale. Only the large firm can afford the plant, machinery and specialists it wants, among whom technical staff are particularly vital in the high-technology industries like telecommunications and aerospace. Though made up of individuals, these enterprises, and especially those of big business, have a macroscopic life, a self-perpetuating urge, of their own. Theirs is a giant's life, with a giant's appetites and foibles. Their cannibalistic struggles, sudden eruptions and equally precipitous downfalls make human existence pale into insignificance. The human being is helpless in their leviathan coils and can be crushed by their convulsions.

Some of them—the world's largest—are listed in the table below.

Table 5.1

The World's Largest Industrial Companies and their Sales in 1970/71

	Sales in 1970/71 financial year, in £million
1. General Motors Corporation	10,123
2. Standard Oil Co. (New Jersey)	6,847
3. Ford Motor	6,148
4. General Electric Co.	3,520
5. Mobil Oil Corporation	3,077
6. International Business Machines	2,999
7. Chrysler	2,938
8. Gulf Oil	2,545
9. Texaco	2,445
10. Royal Dutch Petroleum	2,369
11. Shell Transport and Trading	2,352
12. International Telephone and Telegraph	2,281
13. British Petroleum	2,243
14. Western Electric	2,035
15. United States Steel	2,010
16. Standard Oil of California	1,829
17. Shell Oil	1,762
18. Standard Oil Co. (Indiana)	1,757
19. Volkswagenwerk	1,611
20. Ling-Temco-Vought	1,563
21. Philips	1,518
22. E. I. Du Pont De Nemours	1,513
23. Westinghouse Electric	1,462
24. Unilever NV	1,368
25. General Telephone and Electronics	1,359

Source: *The Times 1000.*

This is not to say that all companies are that large. Small firms—with 500 employees or less—still constitute more than 90 per cent of Britain's manufacturing industry, for instance. In the United States there are more than 300,000 companies, with 500,000 factories, employing 16 million people and producing half the world's manufactured goods. The majority are quite small. In other industrial countries the situation is similar, with firms ranging from the miniscule one-man business to the gigantic, with employees scattered around the globe. However, the giants are playing an increasingly dominant role in determining the shape of international trade and technology. Fiercer competition and growing complexity and expense of equipment is forcing smaller companies into mergers or other links with their own kind, or with government. Though mergers can assist companies to compete effectively in world markets, they almost invariably take their human toll, in redundancies and strained communications. Recent years have seen such rationalization overtake a number of Britain's industries, among them electrical engineering, textiles, aerospace, computers, cement, bricks, bakeries and food processing, and redundancies along with it. Larger enterprises are created, and one of the management's most difficult tasks is to make the employees feel that they belong to the new group. The larger the group, the more elongated its communications. The operator on the shop floor feels remote, out of touch with the firm's plans and progress. He will not be able to see whether his role makes any difference, especially when there are scores, even hundreds, of people with similar roles in his immediate vicinity. He feels neglected and unimportant, and in his mind the company personnel become polarized into 'them'—the bosses—and 'us', him and his fellows.

The fact that groups can become too large and too impersonal is just beginning to sink in. Some plant managers will now admit that they cannot properly involve everyone in company planning when the labour force totals much over 1000, and that further expansion should be achieved by hiving off a subsidiary of similar size. Motivation, it is being recognized rather tardily, is at least as important as money if a person is to be satisfied with his or her job.

These might have been considered elementary problems but ours is really the first generation that is being compelled to think about them in the context of the giant corporation or the mammoth government agency, which are chiefly phenomena of the postwar era. Previous generations had to come to grips with the mass exploitation and squalor of the first Industrial Revolution and it took a long time to do much about them. Now we face plenty of urgent problems bound up with the impact of the present technological revolution on people. These are every bit as serious and will tax all our modern social skills to the full.

6 · Energy: Technology's Lifeblood

Energy supplies are the lifeblood of technology. Without copious quantities of fuels—especially coal, oil and gas—there could be no industrial state, none of the comforts of modern living, and Man would have had no more than a puny claim over his environment. He would still see his corn ground between the stones of the windmill, and his materials fashioned on machines turned by a water-wheel. Power is needed in substantial amounts to drive modern machines in the factory, on the farm, on the construction site: to smelt metals, and then drive the machine-tools that shape those metals; to heat kilns, furnaces and chemical reactor vessels in which useful materials are created out of the earth's raw reserves; to provide heating and lighting for homes, offices, hospitals, shops, garages, schools and so on—in fact for anywhere that Man wants to create an artificial 'micro-environment' in which to live and work.

It may sometimes be difficult to appreciate how much difference it has made to our lives having energy supplies readily available where and when they are wanted. Before that stage men's lives were regulated—much as animals' lives—by Nature's whim. The working day was almost entirely governed by the passage of the sun: without efficient illumination (and if the guttering candle or the lamp is cosy, it is no light to work by), the range of activities that could be carried on after dark or before sunrise were severely limited. Cooking, the provision of warmth, family life, all revolved around the wood or coal fire. The horse's pace determined the rate of overland transport, while the wind drove ships across the sea, leaving them becalmed at its caprice. For thousands of years labourers had to make do with their own muscle power and that of animals, supplemented by the force of the wind and of falling or running water. Fire, which helped them keep warm, bake their pots and their bread, and smelt metals, was no more than a source of heat. Not until the invention of the steam engine was heat converted to power. Since then the situation has altered radically as the solid fuels of antiquity—sticks and leaves, animal dung, wood, coal, peat and possibly pitch—have been complemented by oil (yielding petrol, kerosene, fuel oil, and so forth), natural gas, and finally nuclear power, and more conversion devices (such as the internal-combustion engine) introduced. Technology has a rich choice of different sources of energy at the present time, the deciding factor in the choice being their relative cost. Whatever the choice, we no longer have to sweat and strain over physical tasks as the electric motor or the internal-combustion engine is there to help. We can bask in the artificial radiance of electric light at any hour of night or day; conjure up warmth at the flick of a switch; and, what is more, carry our own power sources (in the shape of electric batteries or petrol engines, or fuel cells) and set them to work wherever we may be.

At one time industrial life could not flourish except around a ready source of energy. A reserve of wood or coal, or some other fuel, drew industry to itself and helped shape the industrial environment. That restriction soon disappeared, however, as Man devised ways of distributing fuels or electricity so that industry could settle where it wished. Now the tentacles of the power industries are everywhere. Lines of electricity pylons, stretching for hundreds or even thousands of kilometres, are a familiar sight of the industrial scene, while pipeline grids for oil and gas, though less stark, are no less valuable in moving large quantities of these fuels smoothly across the world to the places where they are needed. Pipelines may be called upon to bring fuel from some of the remotest places on earth—the frozen wastes of north Alaska, the burning dunes of Africa—and surmount formidable barriers on the way. Pipelines run from gas wells in the North Sea to the British coast (Plate 6.1), under some of the wildest, greyest water in the world; and they climb the massive ramparts of the Alps. In spite of the ever-lengthening networks, higher operating pressures and larger pipe diameters, accidents are rare—thanks to better methods of forming and welding pipelines, of installing and testing them, and of continuously monitoring them in service. With all these ways of distributing energy supplies, technology can penetrate to any point on earth, and even beyond, as space flights have demonstrated.

One of technology's chief characteristics is its appetite for energy. Generally, there is a good correlation between a country's utilization of machines, the amount of energy it uses in relation to its population and its standard of living (see page 34). That is, the more advanced its technology, the greater its consumption of energy per head, and the higher its standard of living. However, the energy can be supplied in various ways, by coal, oil, gas, and so on. In fact, the proportion that each source supplies is constantly

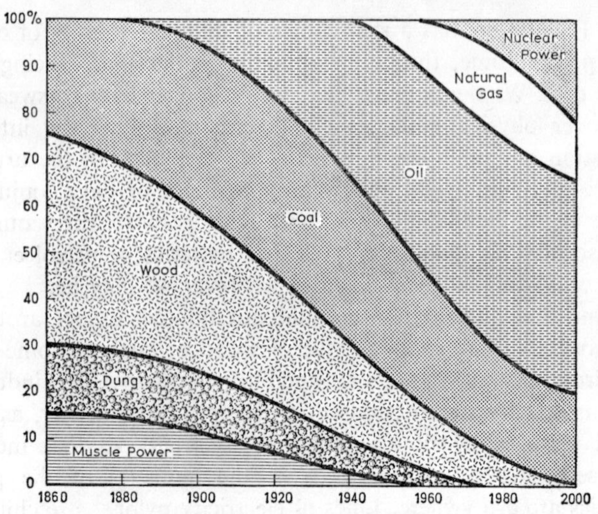

Diagram 6.1 Proportions of the world's supply of primary energy, divided among major sources (Australia and New Zealand Bank)

changing. Diagram 6.1 shows how the overall energy market is shared between various sources, and how these shares have been changing.

The Rise of Oil

The rise of the oil industry has been the most outstanding event of recent technology. Petroleum, in the shape of surface deposits of asphalt, was known and used before 3000 B.C. By A.D. 100 it was being refined (distilled) in the Near East to make naphtha as the base for the weapon 'Greek fire', hurled by catapult at beleaguered fortresses. Although surface seepages of oil were used for liniment and medicine in the nineteenth century, the first successful well drilled specifically for oil was completed only in 1859 at Titusville, Pennsylvania, USA.

The motor car, and later the aeroplane, with their great thirst for fuel, determined the shape and size of the modern oil industry. At first the industry was a supplier of kerosene, lubricating oil and paraffin wax, obtained by distilling the crude oil. The light spirit—gasolene—was thought to be of little value, being too dangerous to burn in lamps, and was often got rid of by burning or running it into rivers.

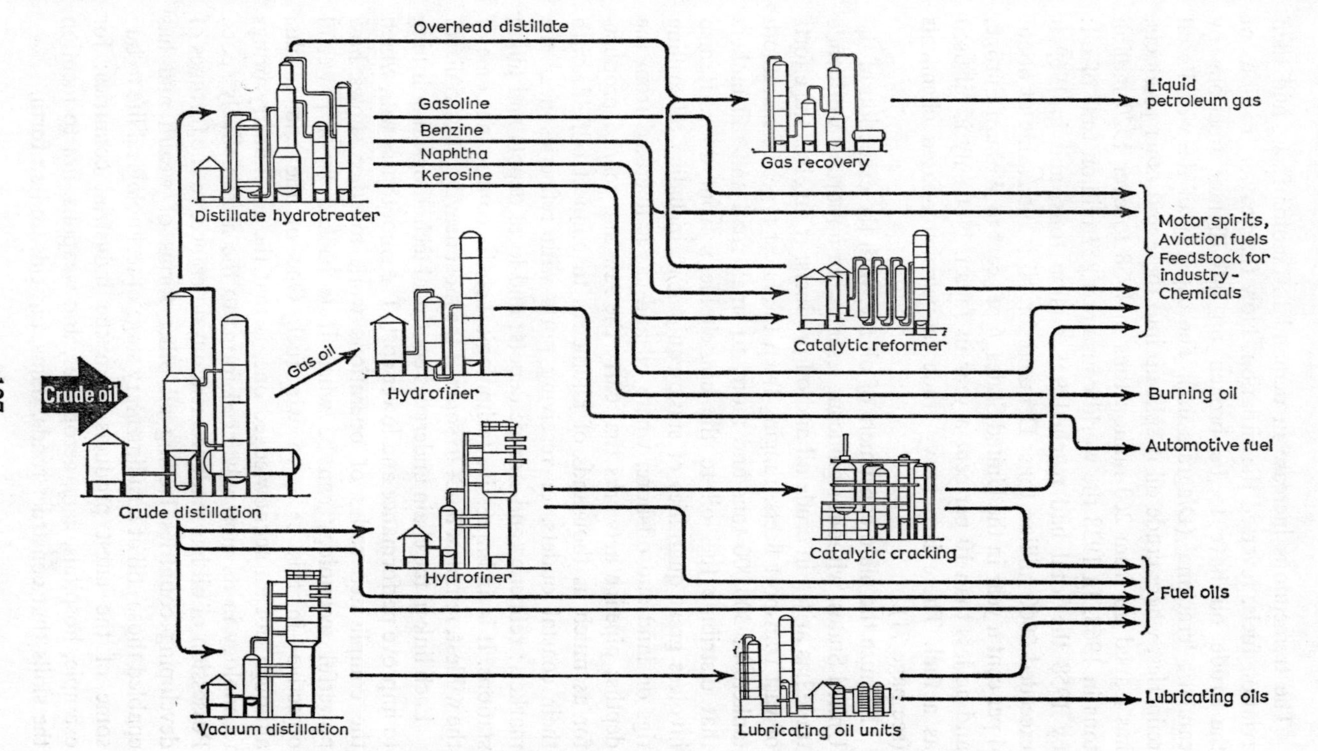

Diagram 6.2 Production of various fractions from crude oil in an oil refinery

The tremendous increase in demand for motor fuels, and then aviation fuels, reversed the situation. Now the heavier fractions of the crude oil have to be broken up into lighter fractions by catalytic 'cracking' (Diagram 6.2). The capacity of the world's oil refineries, where crude oil is split up into these and other products, has leaped up from 260 million tons in 1938 to over 1920 million tons in 1968. In 1938 the world consumed 257 million tons of oil: by 1958 the total had reached 925 million tons, and in 1969 it exceeded 2000 million tons. Demand for oil is increasing at about 4 per cent a year in the United States, 6 per cent in Western Europe, and no less than 10 per cent a year in Japan. Not only is oil used as a fuel. There is a heavy trade in petroleum-based chemicals (see page 71).

Because the main consumers of oil are, with the exception of the United States, without significant reserves, an important worldwide trade has arisen in crude oil and oil products. It involves the efforts of vast transport fleets, ranging from the latest types of mammoth tankers of 300,000 tons and more, to small road and rail tankers that distribute fuel oil to the householder's front door. It also involves great quantities of static equipment, including oil drilling rigs on land and offshore, where oil and gas is liberated from the depths; pipeline networks that carry the raw and refined products for as much as thousands of kilometres to connect oilfields with their coastal outlets, or receiving ports with refineries near the markets; refineries and storage depots; and local depots and filling stations. It is an impressive technological chain, connecting some of the wildest, remotest lands to some of the most heavily industrialized.

Each link in the chain undergoes continual technological polishing to improve performance and lower cost. The smoothness with which the complicated series of operations work together shows how powerful technology can be when it is backed by a powerful objective—in this case, to win fuel. One of the end results, although not an acknowledged one, is that the oil industry brings technology in its most advanced forms to the most unlikely spots. Possession of oil has made tremendous differences to the fortunes of developing countries lacking all other forms of wealth and has enabled those still at a rudimentary stage of technological life to buy some of the latest products from the industrial countries: for example, hospitals, cars, aeroplanes, and weapons, not to mention the skills that construct roads, dams, airfields and so forth.

Oil has been produced cheaply enough to upset coal's traditional dominance of the fuel market. In 1930 solid fuels supplied almost 80 per cent of the world's commercial energy but by 1960 their share was barely half. At present it has dropped to around 40 per cent, and the decline is continuing, as Diagram 6.1 illustrates.

Coal is suffering a fate similar to that of all the previously discovered fuels, of reaching a zenith and then being overtaken by a newer source. It is a familiar story of growth and decay that applies as much to technological life cycles as it does to those of the living world.

It is not only that fluid fuels—oil and gas—are cheap, but that solid fuels are inconvenient to transport and, in their raw state, leave behind ash when burnt. It is difficult and expensive to hew coal out of the ground, wash and sort it, transport it to the place where it is to be burned, then prepare it for burning (in power stations, for instance, which constitute the British coal industry's biggest market, the coal is ground to a fine powder before being blown into the boilers), and finally dispose of the ash, while cleaning grit from the smoke so as to reduce air pollution.

Of course technical improvements can be made at every point. In the latest type of mine, much of the coal digging equipment and the conveyors are remotely controlled from an underground 'nerve centre'. Few men need be at the coal face. Remotely operated machines should be feasible if sensors can be designed to follow the seam of coal as it dips and climbs through the rocks. In that case, completely automated mines would be practicable, and the task of winning coal would become less hazardous. Even without this complete technical achievement, however, the coal industry can fight back in many ways, using machinery to boost the miner's productivity. Most of the coal mined today is cut, loaded and carried about the mine by machines, while wooden pit props are frequently replaced by movable hydraulic roof supports. Above ground, machines sort and wash coal, and feed it into quick-loading, quick-emptying rail wagons to speed its distribution, with less and less human effort as time goes by. The coal can be processed to remove its 'smoky' constituents, and while the smokeless fuel is burnt as a source of heat, these may be used as raw materials for chemicals. When burning ordinary coal, the fluidized-bed type of burner, in which air is blown through a perforated plate into a bed of coal particles, making them behave like a liquid, improves the efficiency of combustion and heat transfer, and reduces air pollution. Alternatively, coal can be chemically treated to yield petrol, though the cost is at present too high. As oil reserves dwindle, though, it seems inevitable that a growing proportion of coal supplies will be converted in this way. But when all is said and done, teams of men still have to slave in the warm, damp, silent darkness underground to wrest coal from the unwilling rock. Until coal can be made to flow, like oil, once the drill has struck, the coal industry's image will suffer on account of this reminder of the grim days of the first Industrial Revolution.

Coal has lost its supremacy to oil and gas and other fuels, and in most of the industrial countries the industry is shrinking. In Britain, it was the largest single employer twenty years ago, with more than 700,000 men on the books. Since then miners have been given generous redundancy terms to leave the industry, or have been moved from old, costly pits whose richest seams have been exhausted to newer ones in the more profitable regions of the East Midlands, South Yorkshire and elsewhere. As the 1960s came to a close pits were being shut down at the rate of one every ten working days. By the mid 1970s there will be perhaps 300 remaining, employing maybe as few as 160,000 men, according to present forecasts. Yet, seen as a whole, world energy demand is growing fast enough to offset this decline. Although coal claims a decreasing share of the total market, actual world coal consumption is rising, decline in one area being countered by birth in another. Coal production is having to be transferred from the old, worked-out seams to new ones, and Britain's situation is a microcosm of the global picture.

However, the fortunes of gas, the third of the chief hydrocarbon fuels, are very much in the ascendant. In 1946 gas met only 15 per cent of the energy demand of the United States, and virtually none otherwise. Today the United States obtains one-third of its energy from natural gas, and world natural gas production, already substantial, is increasing at about 10 per cent a year. In the last few years substantial new reserves have been discovered in the Netherlands, the North Sea, Siberia, Australia, and elsewhere. These discoveries, combined with the perfection of methods for making gas from oil instead of coal, has caused an unprecedented boom in the gas industry, the like of which has not been seen since the richer citizens of London first began to light their houses with gas in the eighteenth century. Besides the ease with which it flows about through pipes, gas has the useful property of shrinking to about 1/600 of its original volume when cooled to the cryogenic region (around −126°C), so that it can be transported liquid in special ocean-going tankers, and stored underground in shrouds of frozen earth—in fact, treated very much like oil.

Electricity occupies an odd place in the energy field. As a man-made power source (generated from other sources, such as coal, oil or gas) it is the most artificial of them all, and thus the nearest to technological ideals, being convenient, versatile, potent yet precise, easily transmitted, and clean. Though used as a source of heat—as the hydrocarbon fuels are—it also acts as a source of light, power, communication and control. Its use is stimulated by industrial growth, technological advance, and by higher ownership of such luxuries as washing machines, hairdryers, electric blankets, mixers and electric shavers.

This incomparable servant of technology and Man begins life in the power station where generators, turned by pressurized water or steam escaping through turbines, form the first link in a chain that finally reaches homes, factories and offices. Diagram 6.3 shows how the power station might be run by oil, coal or nuclear power, as is common in Europe. It might also be driven by water power, as frequently happens in North America, Scandinavia, and parts of Africa, and elsewhere.

From the power station the chain stretches to the customer in his home with his hand on the electric switch. Generally he will not know about technical improvements that are continually being made to the system, nor will he care. All he cares about is having power to hand twenty-four hours a day, every day of the year.

Unfortunately, electricity cannot be stored in any quantity—though batteries are invaluable small-scale reservoirs—and it has therefore to be generated as it is required. The generating capacity of the power supply system has to be carefully matched to the demand, and power stations have to be started up and switched in to the system as demand builds up to its morning and evening peaks, and switched off again as it falls. The distribution system must be reliable and efficient to carry power from where it is generated most cheaply to where it is needed. Such is the technological achievement of control techniques, and the technical excellence of transformers, switchgear and the rest, that one man can control a whole nation's power supplies from one nerve centre, switching power to a

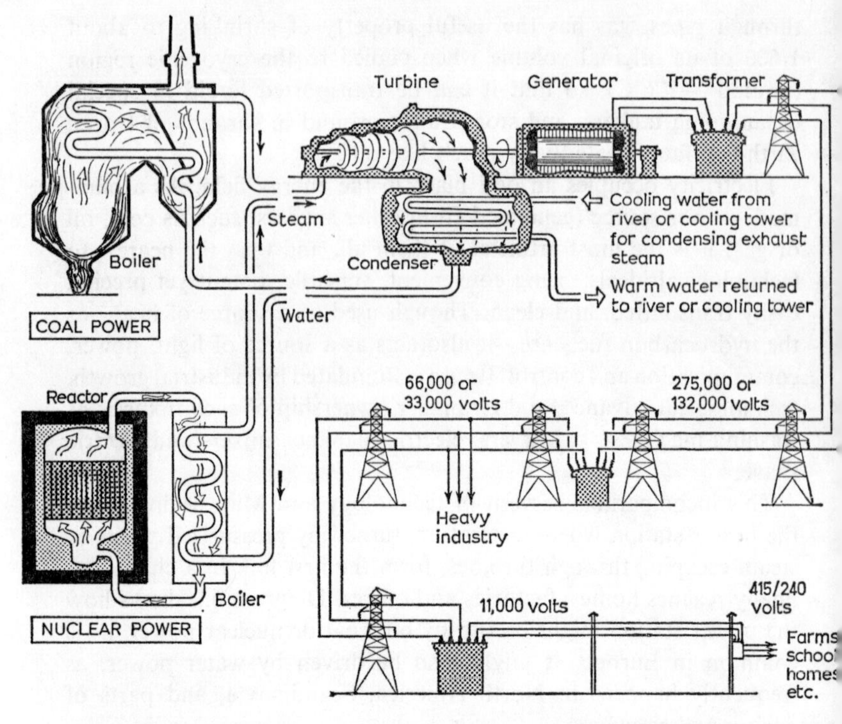

Diagram 6.3 The technological chain linking the power station to the consumer. As is shown, the power source could be fossil fuel—coal or oil, nuclear power, or hydro-power, etc.

region that faces heavy demand from those that are more lightly loaded.

However, in an extensively inter-linked system, a single fault can trigger off a disastrous chain reaction that cripples the whole system. This in fact happened in the great blackout of 1965 in the north-eastern United States, which left an estimated 30 million Americans in darkness. Such a technological disaster—if temporary—serves as a sharp reminder of how much we rely on technology, how much we take it for granted, and how helpless we are without it.

Although that may be the most obvious problem of power supply, it is by no means the only one. There is the unalterable penalty that we always have to pay in using energy. Something is always lost from the energy of the original source along this chain, and the further the eventual user is from the source, the heavier the loss. The best modern power stations recover only about 30 to 40 per cent of the heat in the fuel as useful work. The rest is lost in the intermediate steps—boiling water to make steam, passing the steam through a turbine to drive an electric generator, and overcoming the friction in the machines. A great deal of waste heat (from condensers, where spent steam is condensed) is discharged, usually in the medium of cooling water, into waterways where it can harm fish and other marine life. Cooling towers that discharge heat into the air alleviate the problem.

There are more losses as the electricity is transmitted, but they are reduced by transmitting at higher voltage, though this requires better insulators, more robust switchgear, etc. Attempts are continually being made to reduce all these losses, and to ensure that there are no blackouts (or 'brownouts', as voltage reductions are called). Voltages are being stepped up to increase efficiency of distribution. Not so long ago 69 volts was thought quite sufficient. Now much of the British transmission system operates at 132,000 volts, lines of 275,000 and 400,000 volts carrying most of the heaviest flows. Even higher voltages are being explored. These voltages have to be reduced by transformers before reaching the final user. To get economies of scale in the power station, generating sets are being built in larger sizes. A capacity of 30 megawatts (MW) was thought quite respectable for a British generating set at the time the Second World War ended; the brisk pace of development soon raised that figure. Single sets of 660 MW are being built, larger ones are on the drawing boards, and the 1000 MW unit is not far away.

Some of the waste heat in the power station's cooling water or exhaust gases can be recovered. The 'total energy' approach—a principle of general scope, by no means confined to power stations—seeks to recover heat discarded in one process and use it for another so that one fuel supply meets many needs, whether they are for power, lighting, heating, or whatever it may be. This idea can reach quite a sophisticated pitch in district heating schemes where hot water from a local power station heats houses, schools, and so on.

Another possibility is to burn local refuse—three tons is about equal to one ton of coal in its heating value—in a district incinerator, and use the heat to generate electricity or to provide local district heating. Heat may also be recovered from light fittings, or from warm areas in a building, and used to heat other cooler areas. People naturally exude a good deal of heat which normally goes to waste. Sunshine is another source. Some office blocks already have little conventional heating, and draw their warmth from air circulated round light fittings, and from these other unusual sources.

Although the customer—the ordinary man in the street, or the housewife—may care little for the technical aspects of power supply, they care deeply about matters of amenity. If all other things were equal, it would be best to site a power station near the area it was to serve, in order to reduce the losses of transmission. However, other things are not equal. The question of amenity is often as important as economic convenience, and in the case of a nuclear power station safety is one of the chief considerations. In the early days of nuclear power it was deemed prudent to keep power stations away from large towns or cities because of the radiation risks if there were an accident, and because their voluminous thirst for cooling water meant they had to be sited near the coast or an estuary. Now, it is technically feasible to site reactors safely in the middle of the largest cities and to ensure that the chances of an accident are so remote as to be almost negligible. Yet the man in the street is suspicious. There can be few people who would relish having a reactor at the end of their garden. Then there is the vexed matter of pylons, a blot on many a landscape. Despite technical advances, it is still many times more expensive to put cables underground, largely because they have to be insulated (aloft, the naked cables are insulated by the air) and the heat that is generated as the current flows through them has to be conducted away. In some places, arguments based on amenity gain sway and cables are buried, the extra cost being borne ultimately by the consumer.

These, and other, environmental factors are coming increasingly to the fore in public discussion of the siting and operation of all kinds of technological plant and machinery. They are now among the most important considerations in determining where power stations, especially, are built. As we use power more abundantly and more widely, we have to accept the consequences or pay dearly.

Power from the Atom

Nuclear power, though still in its infancy after twenty years of development, is just starting to tilt the energy balance. Already it is producing electricity as cheaply as coal and oil, and by the year 2000 it will probably supply a large part of the world's power. It is a direct result of the wartime atomic bomb programme and, in particular, of the massive Project Manhattan—an American initiative that must count as one of the wonders of the modern world. It harnessed £700 million and the labours of 600,000 men and women to turn the promise of the first apprehensive experiment of the atomic age—when on 2nd December, 1942, the first atomic pile went critical in a converted squash court under the stands of Stagg's Field, Chicago—into a grim reality only three years later, when the first atomic weapons in Man's history exploded over Hiroshima and Nagasaki. The Stagg's Field experiment demonstrated that the chain reaction of nuclear fission could be initiated and controlled, and so laid the foundation for both atomic weapons and atomic power. The explosion of the atomic bombs showed that uncontrolled nuclear fission is indeed an awesome force.

It took longer to demonstrate that controlled nuclear power was practicable. As warfare subsided, there were teams in North America, Europe and the Soviet Union with sufficient knowledge of the hazardous new techniques of handling atomic fuels and atomic reactions to apply to peaceful power generation. But there were also formidable problems involved. In 1954 the Russians managed to get a 5 MW power reactor working. In 1956 the British reactor at Calder Hall, Cumberland, that had been built to make plutonium for bombs, started to supply power to the electrical grid. The United States launched its first nuclear-powered submarine, *Nautilus,* in 1954, and brought its 60 MW land-based Shippingport reactor into operation in 1957.

Since then, vast amounts of money and manpower have been devoted to the nuclear power business, and a commercial warfare has broken out between the rival reactor systems. There are two chief types, the gas-cooled, typified by the British systems—derived from the Calder Hall pile—and the water-cooled type favoured in the United States. The difference lies in the method by which heat generated by nuclear fission in the fuel is conducted away. In a gas-cooled reactor the gas—often carbon dioxide—is blown through the reactor core around the fuel by large fans. The hot gas passes through boilers where it gives up its heat to raise steam, which then drives turbogenerators in the usual way. The cooled gas is recirculated through the reactor. Britain, the first country to adopt nuclear power on a major scale, has based its nuclear programme on reactors of this type, though it is still a matter of contention whether they generate electricity as cheaply as oil or coal-fired stations (Plate 6.2). However, the reactor is more efficient when it is run hotter, and although this imposes new problems of design, thermal stability, materials, etc., good progress is being made on these high-temperature reactors. It is planned to generate 25 per cent of Britain's electricity from nuclear power by about 1985, compared with around 12 per cent at present.

The United States followed up its early interest in the water-cooled types, where water fulfils the same job as the gas does in the previous reactor type. The water may be allowed to boil, in which case the steam can drive a turbine directly; or it may be pressurized to stop it boiling, in which case the super-heated water has to pass through a boiler where it gives up its excess heat to a separate water system: there, steam is again raised and piped to a turbine as before. Not only have these been ordered widely in the United States, since the original Shippingport model appeared, but they have also been ordered by many other countries across the world. If in the United States nuclear stations supplied less than 1 per cent of its electricity in 1968, the proportion is expected to rise sharply to about 25 per cent by 1980.

Ever since the early days a great number of alternative reactor types have been studied, and in many instances experimental and prototype units have been built, involving a variety of permutations of coolant, fuel, and moderator (the substance in the reactor that slows down neutrons released by the nuclear fission so that they can effectively react with other atoms in the fuel). The snag is that by and large these reactors make use of only a few per cent of the uranium that is dug out of the ground for power generation.

The next generation of nuclear power stations are of a different type—the fast breeder. This reactor does not have a moderator (hence the neutrons are 'fast') and it can produce—or 'breed'—plutonium from uranium that could not otherwise be used in reactors. It means that the fast breeder is expected to make available up to 75 per cent of all the energy in the uranium—or in thorium, another nuclear fuel. Plutonium is itself a fuel and a nuclear 'explosive', and a breeder can make more plutonium than it consumes as fuel. All industrialized countries are planning to use fast breeders, cooled by liquid sodium, as the foundation for the next round of nuclear power producers, say in the late 1970s or early 1980s. An experimental unit of the type has been operating well at Dounreay on the north coast of Scotland since 1959 and a prototype power station is being put into operation there, based on this type of reactor.

Whatever hesitations there may be about the present economics of nuclear power generation, there are none about nuclear reactors being economic power sources in the near future. Many types are being transferred from experiment to power generation and other operations, such as the creation of radio-isotopes. Thus in 1962, twenty years after the first chain reaction, there were more than 400 reactors in the world, though most of them were low-power units built for research. In contrast, of the 480 or so reactors in operation at the beginning of 1970 rather more than 100 were power reactors whose prime function was to produce electricity. The world's nuclear generating capacity is expected to grow from 25,000 MW in 1970 to 300,000 MW in 1980.

These nuclear generating stations will require large quantities of fuel, and a vast new industry is springing up to meet their needs. Its business lies in mining ore, concentrating the metal and making fuel elements and reprocessing spent fuel. In time it will probably come to rival and then surpass the industries built up around coal, and even oil. Its turnover, which reached about £100 million in 1970, is confidently expected to reach £1000 million by 1980 and go on climbing.

Any state aiming to set up in nuclear power or nuclear weapons must be able to obtain the fuels or fissile substances it needs to produce the chain reaction; and it must have the means to handle radioactive materials safely while they are in use, and dispose of them when they are spent. The first task generally involves the enrichment of natural uranium (U–238) in its much rarer form, U–235. The customary means—adopted to prepare uranium for the atomic bomb—is gaseous diffusion, a process that entails thousands of repetitive steps of 'filtering' corrosive uranium hexafluoride gas through membranes, and then reclaiming enriched metal from the treated gas. The necessary plant is large and costly, and consumes colossal amounts of electricity: a 'small' plant can occupy a building a kilometre or more long, while a single plant of the type in operation in the United States can consume 2500 MW, enough power for three cities the size of Birmingham.

The process is now being supplemented by the ultracentrifuge method. Here the gas is spun in very high-speed centrifuges, flinging the heavier U–238 outwards and leaving the inner gas richer in U–235. The method uses less power, and the units do not have to be immense to be economic. Hence it would be easier for a developing country to employ this than the gaseous diffusion method, which is in any case restricted to plants in the United States, Britain, France, Japan, the Soviet Union, and China. Fears have been expressed that the new method would open a low-cost route to nuclear capability, and that unscrupulous rulers of emergent states might seize on it to claim that glittering prize. The idea of a number of states in, for instance, the Middle East being armed with nuclear weapons is indeed a daunting one, and one that the nuclear non-proliferation treaty does little to dispel.

The reprocessing of spent fuel elements, and the safe disposal both of the radioactive waste products removed from them and of the active materials created during all phases of the nuclear business, is a most serious matter for all states. A number of countries are building reprocessing plants to handle the predicted volumes of spent fuel from nuclear power plants, but so far only a few have felt the strain of disposing of the large amounts of waste involved (see page 147). Waste disposal is bound to limit the rate at which nuclear power is introduced.

An important offshoot of the nuclear supply business that thrives largely untroubled by such considerations is the trade in radioisotopes. For example, the worldwide industry for sterilizing medical equipment (e.g. syringes) with radioactive sources is worth several million dollars annually. Either sources such as cobalt 60 or accelerators that produce beams of equivalent radiation are used to diagnose and treat cancer, thyroid diseases and other ailments. Food sterilized by radiation is on sale in many shops. Radiation treatment is being adopted in the creation of better wheat, barley and rice through radiation-induced mutations; in the 'curing' of paints or plastics and the vulcanization of rubber; and in the treatment of sewage. The release of insects that have been radiation-sterilized so that they produce no offspring on mating is a pest control method which has proved effective without leaving any harmful chemical residues. Industrially, nuclear gauges employ the penetrating properties of radiation to determine the structure, thickness, moisture content and other properties of materials. Then again, a small nuclear generator powered by a radioactive source was the sole supply of electric power for instruments left on the moon by Apollo 12. While related devices have been designed to power implanted pacemakers that keep unsteady hearts beating in correct time. In these and many other small ways nuclear fission is as useful to Man as it is in the mighty nuclear reactors.

Power without Pollution

The basic drawback about using all these forms of energy is that they pollute the environment, though some pollute worse than others. Coal is the worst. A large coal-fired power station of 350 MW emits each day about 75 tons of sulphur dioxide—a heavy, colourless, corrosive gas—16 tons of nitrogen oxides (another potentially harmful gas) and 5 tons of ash particles. That is not to say that all are discharged, of course. Modern power stations have efficient extractors that collect the grit from the exhaust fumes and a few have plants to remove sulphur dioxide, but these are comparatively recent additions. The fluid fuels, though they do not give rise to much grit—if any—do liberate combustion gases often containing large amounts of sulphur dioxide. Car and lorry engines emit, among other things, unburnt petrol or diesel fuel, sulphur dioxide, carbon monoxide (a deadly poisonous gas), oxides of nitrogen, and traces of compounds added to the fuel to give it anti-knock or other desirable properties. Besides all fuels liberate water vapour and carbon dioxide when they burn. Although the amounts involved in burning one match can hardly be said to pollute the air, the vast quantities liberated by the wholesale combustion of all the fuels needed to keep technology going may actually be enough to upset the overall balance of the earth's entire atmosphere (see page 266).

Then again, there is thermal pollution—the liberation of waste heat to the environment, where it may have harmful effects. This is becoming a common problem, especially on inland waterways that have to bear the heated discharges from several power stations, since each one raises the temperature of the water a little further in the process of condensing the steam from its turbines back into water that can be fed into the boilers again. The cumulative effect can be drastic. Nuclear power stations, which share in this thermal pollution, harbour their own form of pollution—radiation—which, without the strictest controls could wreak even worse harm. So far, the progress of nuclear power has been undisturbed by any serious accidental release of radioactive substances. The risk remains.

With public concern about pollution well and truly aroused, the search for power devices that do not produce so much debris has intensified. One non-polluting source is of course the sun, the original source of most forms of earthly energy, and the vital one that sustains all life. Each year the earth receives from it something like 100,000 times more energy than is produced by fossil fuels. One per cent of that is absorbed by plants and converted through photosynthesis into vegetation. In this continuous creation of vegetable matter we have a constantly renewed supply of solar energy, which, as we saw, has been utilized (in the shape of wood and vegetable waste) for many thousands of years. If instead of being burnt, the vegetation rots, the heat is given out more slowly, though only rarely—as when a rotting hayrick catches fire—do we become aware of it. In Britain sunlight is hardly a practicable source of energy but other countries are blessed with better weather and sunlight is a feasible source. Even so, it is very diffuse and must be collected and concentrated, as in a solar furnace. In the French Pyrenees, for instance, experimental furnaces that attain temperatures of 3000°C or more, enough to melt most metals, are employed to test the flash effects of atomic bombs, to produce exotic substances such as high-purity quartz for the electronics industry, and for other purposes. Although major industrial applications are still far off, solar furnaces can also do a variety of useful jobs. Solar stills produce drinking water from saline water in such sunny lands as Greece, Australia and Portugal. Solar furnaces can convert solar heat into mechanical energy by driving engines running on hot air or steam, and there are a multitude of consumer gadgets, including solar-powered radios, fans, cookers, water heaters, and refrigerators. In space, sunlight is one of the prime power sources (the others being nuclear power and fuel cells), and space satellites obtain electricity from silicon cells illuminated by the sun.

Then there is geothermal energy, inside the earth, which can be harnessed in those regions where there are volcanoes, hot springs, geysers, and the like—where heat from the earth's core raises the temperature of subsurface water. As long ago as 1904 Italian engineers had employed the power of steam from underground to drive turbogenerators to provide electricity, enough now to drive most of Italy's railways. In Iceland, New Zealand, the Soviet Union and Japan, hot water welling up from underground can be piped for district heating, or the steam can be used for heating or to drive turbines. In western Siberia, Soviet geologists searching for oil and gas discovered an enormous underground hot-water basin, larger than the Mediterranean. The water temperature reaches as much as 160°C deep down. The waters have been injected into the area's oil and gas reserves to force more of the fuels to the surface. Because they contain various beneficial salts, the waters are also good therapeutic agents, and numerous health centres draw on these natural supplies. There are small-scale projects in a number of places in the Soviet Union.

Tidal power is another natural source. A barrage built across a coastal bay can trap water when the tide is high, and then release the water through turbines at low tide. The turbines drive generators to provide electricity. The idea has been put into practice on the Rance estuary in France, and the Soviet Union is planning to construct two powerful tidal power stations on the White Sea, drawing on the experience gained from the operation of a small experimental station, with a capacity of several hundred kilowatts, which has been running at the mouth of Kislaya Bay in the Barents Sea for over a year. Similar schemes have been proposed for various British sites, including Morecambe Bay and the Wash. It has even been suggested that the surge of water through the English Channel could be harnessed in this fashion.

Energy Converters

No fuel is any use without some device to convert it into the power that technology can use. Normally this conversion is effected by burning the fuel in air; thus a petrol engine burns petrol and turns the liberated heat into mechanical energy through the intermediate of the piston moving in a cylinder under the thrust of the hot, expanding combustion gases. One of these converters, the turbine, is a common power source in power station, factory, aircraft, and ship.

However, though their basic shape and function may be familiar, their detailed design and construction change continually under the hand of the technologist, allowing him to extract still greater performance from a given weight of engine, often for a lower fuel consumption. So that advances in materials technology can radically affect engine performance. Rapid improvements in nickel alloys used in jet engines have allowed the engines to run hotter—maximum operating temperatures have risen from 800°C to 950°C in the last decade—and so be more efficient.

Designers can also wring more power from a given weight of engine for land-based vehicles, and again aluminium and other light alloys play their part in replacing heavier metals such as steel and cast iron.

But it is also possible to make more fundamental changes, occasionally. Thus under critical scrutiny the reciprocating internal combustion engine (the petrol or diesel engine) proves to be an awkward, complicated thing, that surely owes its lineage to the reciprocating piston-and-cylinder arrangement of the early steam engine. Designers are just starting to employ the simpler rotary engine instead, and while the gas turbine appears to be the more suitable version for heavy vehicles, the car seems to need another type of rotary, of which the Wankel is probably the best known (see Diagram 6.4). Here the reciprocating pistons, crankshaft, valves and so forth are replaced by a smoothly rotating shaft with three lobes. If it is much more compact, it nevertheless has associated problems. The seals at the tips of each lobe wear quickly as they rub across the inside of the cylinder. In addition, the shape of the combustion chamber makes for a dirtier exhaust than that of the best piston engines; and this

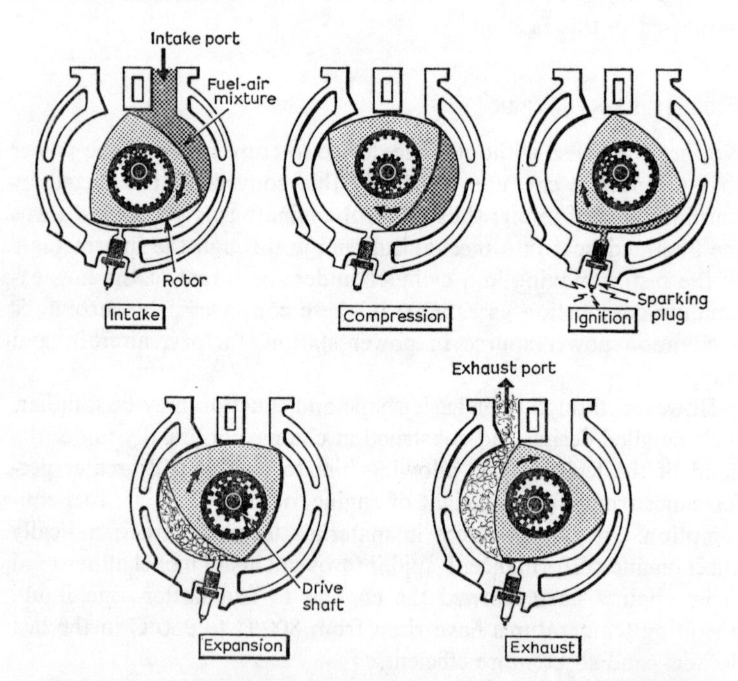

Diagram 6.4 Rotary piston-engine sequence. Air is drawn in with the fuel, and the mixture is compressed, ignited and explodes, finally being forced out by the rotor. The engine is much simpler in design than reciprocating types

may have to be overcome perhaps by pre-heating the air, or cleaning up the exhaust, or other means. If these matters are solved, then it is likely that this kind of rotary engine will assume much higher importance as a power unit for cars, and that the gas turbine will take over from diesel engines where powers of 350 brake horsepower or so are needed for the long-distance haulage of goods and passengers. Much development work is being done along such lines in the United States and Europe. Other factors, like public concern over air pollution, affect both design and performance. For instance, the requirement that cars be fitted with anti-pollution units, and that aircraft engines be fitted with noise and pollution suppressors, materially alter weight, initial cost, and operating cost of the engines, quite apart from their performance.

The fuel cell is something of a different kind, as it converts chemical energy directly into electrical energy, like a battery, with the aid of a catalyst. There are no intermediate moving parts, and the reaction can go smoothly in solution at low temperatures. Unlike a battery, however, it draws the energy from a fuel that is piped into the cell. In terms of efficiency it may be much better than the other converters: a typical car engine has an efficiency of about 20 per cent; a large power station about 40 per cent; and a hydrogen fuel cell between about 50 to 80 per cent. Until about seven years ago most fuel cells used hydrogen as a fuel, because it was simple, although more expensive than the common hydrocarbons. But because of the cost, hydrogen cells are most applicable as power sources for spacecraft and portable military communication systems, and a few other specialized uses. Three fuel cells 'burning' oxygen and hydrogen supplied power and water for the Apollo 11 astronauts on their journeys to and from the moon. Now the search is on for devices that can take cheaper fuels and are themselves cheap and reliable.

The energy converters we have considered so far burn a fuel in the oxygen of the air, or in their own special supply. In space, however, there is no oxygen to burn, and so space vehicles have to carry their own oxygen, or a substance that performs a similar role, to support the combustion of the fuel. Another possibility is to dispense with a combustible fuel altogether and accelerate charged particles—ions—in a stream from the back of the rocket, their reacting thrust on the rocket engine pushing the rocket in the opposite direction. This ion engine would be employed for powering a rocket to go into the far reaches of the solar system.

Power for the Future

Energy is vital for our spreading technology. The world's energy consumption rises steadily, as populations increase and people become materially better off, and the web of technology more entwined with life. Each year we use 5 per cent more primary energy, in the form of coal, oil, gas, and other fuels, than we did the year before, and about 8 per cent more electricity. Though the rate of increase in the richer countries may fall off in the future, as technology satisfies their material wants, it will probably be balanced by a rise in the poor countries, if the current trends in international aid are maintained. Looking into the future, it is forecast that energy consumption in the year 2000 will be about four times that in 1970.

However, the increase cannot continue. The fuel reserves cannot stand it. Man is overtaking the supplies of coal and oil, subterranean stores of energy inherited from perhaps 200 million or 300 million years ago, and in a few centuries—or only another hour or two on our imagined time-scale—they will be exhausted. This has become clear only recently, with the rapid increase in our demands on these irreplaceable supplies. Man's consumption of the hydrocarbon fuels hardly made any impact before the nineteenth century, but in that century alone Man used almost half the total used in the preceding nineteen centuries. In the next thirty years Man will consume as much energy as he has since first appearing on earth.

There is the case of oil. The great growth of the oil industry has been one of the landmarks of twentieth-century technology. World oil production is currently around 2000 million tons a year, while presently proved reserves stand at about 70,000 million tons, excluding some very large deposits in shale and tar sands in, for example, North America and the Soviet Union. The oil industry currently extracts each year less than 10 per cent of reserves and proves more reserves all the time. New fields are being discovered continuously, for example in Africa, South America, the North Sea, North Alaska, and Australia. Yet the easier oil fields have already been tapped and new ones are becoming more expensive and more difficult to find and exploit. The search is having to go further afield, into more hostile areas closer to the poles. It is true that we shall be granted some respite by drawing for a time on such vast deposits as the Athabasca tar sands in northern Alberta (Canada) which extend in a 60 m (200 foot) thick layer containing an estimated 50,000 million tons of oil. Then there are large reserves in the oil shales of Colorado. Both lie for the most part on the surface. Each one is believed to contain as much oil as the present proved reserves of the Middle East. At present, Middle East oil is cheaper, because the sand or shale has to be heated to drive off the crude oil, but as oil becomes scarcer, and more expensive, these and similar sources, if there are others, will become economic. For a time they, too, will yield world supplies. Nonetheless, however good we make the technology for finding and exploiting oil reserves of all kinds, there will come a time when the oilfields and the gas fields run dry, or when it would be prohibitively expensive to extract the fuels left in them. Coal and related fossil fuels, such as peat and lignite, will probably last longer—perhaps 400 years—but they too are limited.

Other sources are needed. Some countries are able to utilize geothermal, solar, tidal and other forms of non-polluting power. Many cannot. They, and indeed the world in general, are now looking to nuclear power. Uranium, its chief fuel, is a much more 'concentrated' fuel than coal: 1 kg of uranium can perform as much work for technology as almost 3 million kg of coal. At present our reserves of uranium are almost untouched, although extensive mining and processing operations have been mounted since the early 1960s to extract nuclear fuel from the natural ores. Even so, the uranium will only last for a limited time. Experts believe that the free world's known reserves will be committed by the early 1980s.

At first, nuclear power will be confined to the industrialized nations of North America, Western Europe, the Soviet Union and Japan, and indeed only about 2 per cent of the world's capacity is at the moment installed outside these countries. But very soon the developing countries will want to start up their own reactors. These countries will find encouragement and ready practical support from the industrial nations, which are only too ready to sell their various systems to outside customers to help recoup some of the heavy research and development costs. The handicap is usually the developing countries' lack of money to pay for this diffusion of advanced technology, but loans and other forms of aid are generally devised to take care of this.

Of course, as ores become scarcer their prices will rise. The alternative of extracting uranium from seawater is far too costly with present methods. Although nuclear power may be economically competitive with fossil-fuelled power at the moment, its future—like that of the other fuels—will be governed by the fuel supply, and only time will tell which set of costs will rise faster. The position will ease temporarily as the fast breeder comes into operation in the late 1970s, because it makes use of a much greater proportion of the naturally occurring uranium. Other fuels, including the naturally occurring thorium and the man-made plutonium, will be other options. But even they cannot last for long if the world's energy demands continues to increase.

Perhaps the only long-term hope is the fusion reactor, fuelled by 'heavy water', which exists in vast quantities in the oceans. Present nuclear power stations extract their power from nuclear fission, that is, splitting of heavy atoms into smaller parts with liberation of energy. It is also possible to extract energy by fusing light atoms together. It is done in the H-bomb, and is also the process that takes place in the sun to give us light and heat. Needless to say, it is an extremely hazardous and troublesome process to control, as the temperatures at which it occurs are measured in millions of degrees, and the gas in which the reaction is triggered has to be held at that sort of temperature inside a 'bottle' of magnetic lines of force away from any solid materials, which would instantly vaporize on contact. Although the fusion reactor is being studied in most industrial countries, it is still proving a recalcitrant thing to handle, and some time—perhaps twenty years—will elapse before it can be tamed.

Radiation is another problem with nuclear power. Its effect is not immediately apparent, and may work insidiously. The dangers of exposure of individuals to large doses of radiation are well enough known. The long-term exposure of whole populations to small doses is not. All workers in laboratories and other establishments dealing with radioactive substances have to be protected from the radiation by massive shielding of steel, lead, concrete or other impervious materials, and the time that they spend in proximity to such substances is carefully limited (Plate 6.3). They and their clothing are monitored to make sure that no radioactive dust is clinging to them when they leave work. Reactors themselves are shrouded in massive shields of steel and concrete, and are continuously monitored by extremely sensitive instruments for any signs of leaks or excess radioactivity. There are elaborate safety procedures which instantly shut down the reactor if a fault is detected. Yet the population at large is not protected, except by the skill of the designers and the constructors of the nuclear plants, and there must always be fears that if an accident did occur—and the chances are extremely remote—large numbers of people would be exposed to danger.

Moreover, in its lifetime the nuclear power station, although it produces virtually no air pollution (since there is no fuel burned in the air), does produce radioactive wastes of various levels of activity. Low-level waste is usually diluted in large volumes of air or water and returned to the environment. Medium-level wastes are mostly sealed into containers—steel and concrete drums, say—and either dumped in ocean depths, with the hope that they will leak out gradually over a long period in the future and become diluted to harmless levels, or buried in disused quarries, worked-out salt mines, or other sites.

The most intensely radioactive wastes—byproducts of the fission process in the reactor which are removed at the reprocessing plant—have to be stored on land in tanks of steel and concrete, or in vaults that have to be cooled and continually checked for leakage. Some of the products decay so slowly that it will be at least twenty years before anything can be done with them. The prospects of a world whose energy needs were met by nuclear power are indeed daunting. To supply the power required in the year 2000, it would require 10,000 even of the fast breeder reactors, each of 3000 MW, together burning 10,000 tons of uranium or thorium a year. They would produce large quantities of waste products, and the question of where to store the accumulated hot wastes from these reactors is an intractable one.

The reactors themselves gradually become more radioactive. Active substances accumulate inside the reactor and after perhaps twenty-five years or thirty years it has to be abandoned, or dismantled so that the 'hot' parts can be removed by remote-handling techniques for storage. Perhaps the steel and concrete shell can be covered with earth. No one really knows what will happen to them. None have been in operation for long enough.

The public outcry over the dangers of pollution may become so strident that in the long run people will choose to restrict nuclear power and the dissemination of nuclear materials, and deny themselves its benefits rather than risk its attendant evils. They may go further and, on the same grounds, object to the continued use of other fuels.

We began this chapter by emphasizing that technology needs energy. We now see that the consequences of using energy are far more serious than could have been foreseen at the time of the first Industrial Revolution, or at the beginning of the nuclear power programme. In fact, we face a dilemma. If we continue to burn fuels in these prodigious quantities by present inefficient processes we shall cause massive pollution and exhaust our fuel reserves before much longer in the development of mankind. In addition, the heat liberated from the burning of these fuels may raise the temperature of the earth with unpredictable and probably fundamental disturbance of the climate (see p. 266).

If, on the other hand, we deliberately restrict our use of fuels in order to conserve resources and reduce pollution, we shall not be able to support technological progress at its present rate, let alone try to increase it. Our material standard of living will suffer. The good technology and the bad will grind to a standstill if the trend goes far enough, and without strict control by an enlightened government the socially worthwhile technology will be abandoned first in favour of projects geared to exploiting existing resources, or the winning of new ones—projects that have a strong military flavour.

Like a deficiency of materials, an energy deficiency will prevent the fulfilment of our expectations of a continually rising standard of living. We shall, in a sense, have to stop growing. What will happen then? Will men have to go back to living in the warm equatorial zones? Will all the complex paraphernalia of life elsewhere be abandoned, as a barren technological tundra of steel and concrete while the remnants of its civilized populations settle in the warmer zones and burn sticks, as their ancestors did? Is that to be the end of the glorious revolution introduced with the steam engine? Even if it never comes to that, a restriction of energy supplies would fall hardest on those states that now consume the most energy per head of population, that is to say, the industrialized countries. In 1960, the United States with only 6 per cent of the world's population consumed over one-third of the world's commercial energy supplies, whereas India with almost 15 per cent of world population used only about 1·5 per cent. Much of India's energy comes in fact from non-commercial sources such as wood, animal dung and animal power. Will there be a moment when the United States will also have to turn to such sources?

Unless a radically new power source—such as the fusion reactor

—comes to the rescue in time, technology will undergo a progressive collapse from energy starvation within a few centuries.

Table 6.1

World Fuel Reserves and Fuel Consumption

Type of fuel	*Present annual world production (million tons coal equivalent)*	*World's recoverable reserves (thousand million tons of coal, or equivalent)*
Coal and lignite	3,000	7,600
Petroleum	3,000	408
Natural gas	1,300	323
Tar-sand oil	little	61
Shale oil	little	39

An estimate of the world's recoverable reserves of fossil fuels, compared with the rate at which they are being extracted, in terms of the approximate annual production. All figures are in millions of metric tons of coal equivalent, i.e. the quantities of each fuel that produce the same amount of heat when burned as a million tons of coal (first column) or as a thousand million tons of coal (second column).

7 · The Explosion of Communications

Technology has released a great flood of communications. It is as though all mankind had been queueing up, impatiently waiting for the inventions that would enable them to talk across the world. Before the nineteenth century communications were primitive. For thousands of years messages could travel only at the pace of a living agent—a runner, a horse, or a carrier pigeon—and although many tried to improve the speed and range through such devices as signal fires and drums they made little difference. The invention of ways to harness electron flows to carry messages in the telegraph and telephone meant that men could communicate with the speed of light, and could send messages over the horizon. The development of these two inventions and of the undersea cable, the camera and the gramophone all played their parts in releasing the floodgates. When radio was discovered and put to use soon afterwards the torrent flowed more fiercely. Now telephone cables, microwave networks and radio beams crisscross the globe, enabling governments, business organizations, individuals, and even machines to talk to each other almost instantaneously across continents and from one continent to another.

Over the past five years world telecommunications traffic—including television, radio and telephone signals, and data link services—has doubled, and its growth is expected to accelerate in the next five. The growth of data transmission, by which computers send information to each other, and by which all kinds of business and official bodies transmit facts and figures one to another, is much more rapid than that of speech transmission. In the next five years alone thousands of millions of pounds will need to be invested in new equipment to cater for the demand, and many new devices will have to be perfected to carry the potential load.

We can gauge the rapidity of this growth by watching the progress of a common communications device, the humble telephone. Diagram 7.1 shows how swiftly it has been adopted throughout the world.

The accumulation of information is almost as rapid. The amount of accumulated knowledge in the world is doubling every fifteen years or less, as one generation hands on its recorded findings to the

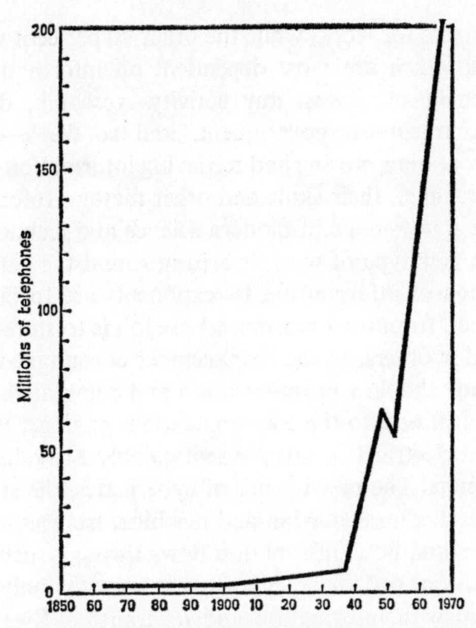

Diagram 7.1 The number of telephones installed throughout the world

next. In scientific fields the amount of knowledge acquired since Man's origins up to 1950 has more than doubled since. Ninety per cent of all scientists who ever lived are alive today, and they are publishing more than 2 million articles a year—chiefly in an estimated 30,000 scientific journals, whose number has tended to double every fifteen years. Before the year 2020, when children now at primary school will be starting their last decade of work before retirement, the present volume of human knowledge is expected to multiply at least tenfold.

The character of the world's major industries is changing (see page 22) from those concerned with things to those concerned with information. The 'information industries'—involved in producing, storing, retrieving and presenting information—are becoming prominent in the world economy. Among them are telecommunications, photography (including microfilm), the computer industry, and xerography and other reproduction processes. Companies in these fields have been able to enjoy growth rates as high as 25 per cent a year.

At present, production takes less than half the world's work force; in less than thirty years all production could be done by 10 per cent of those available for work, while the other 90 per cent will be in the service areas, which are most dependent on information. Already the management of almost any activity—research, development, manufacture, transport, government, and so forth—depends on gathering, processing, storing and retrieving information just as much as on the personnel, their skills and other factors. Information processing is the *sine qua non* of modern science and technology.

Already, a new type of work is arising round the generation and communication of information. Its exponents are to be found in a range of places, from the librarian, whose job is to store and retrieve information for others, to the management consultant whose prime task is to study the flow of information and commands through the largest organizations, to the communications engineer who is trying to ensure that electrical signals are sent quickly and reliably through electrical circuits. The new science of cybernetics, the study of communication and control in Man and machine, has sprung up to help people understand how information flows through such a hierarchy and how decisions are taken. But these are not the only people who are concerned with information and its transfer. Everyone is: the office girl who is eternally typing away at letters and memos, the switchboard girl trying to handle all the phone calls, and the top manager who is trying to sift out what is important, vital and urgent from all the rest of the data supplied to him by subordinates. The man in the street, too, is—or could well be—swamped by the great choice of papers, magazines, radio and TV programmes, films and other media that impinge on him.

We cannot handle more than a tiny fraction of all the information that is available to us. This is another radical difference from earlier ages. When the Royal Society of learned scholars was founded in England in 1660, one well-educated man could know about, if not know, all there was to be known concerning the world in literature, science, arts and other fields. Now, although a schoolboy is probably better informed than the typical seventeenth-century scholar, it is impossible for one man to know more than a fraction of what the human race has so diligently collected in the meantime. Ours is the most heavily documented age in history, and the world's libraries and filing cabinets are already full to overflowing, yet the relentless flood pours forth faster every year.

In all fields of communication, then, which includes not only storing and retrieving information but also transmitting it, technology is being geared to sending more messages over a given communications channel and storing more information in a given space. Let us look at some aspects of this work in more detail.

The Crowded Spectrum

Nowhere is the explosion better exemplified than in the task of sending messages across the world by broadcasting and by cables. Take broadcasting first. The 'air'—that is, the electromagnetic spectrum (see page 43)—becomes continually more crowded as one broadcasting operator after another applies for communication channels on which to transmit.

Each new television channel, for instance, requires the equivalent of 1000 voice channels, to transmit all the information describing the picture and sound. Colour television makes further demands because information about colour also has to be transmitted. Again, each new high-speed link for sending data from one machine to another (teleprinters, for instance) takes the equivalent of about 250 voice channels. To the growing babble of existing users, such as the military, police, fire services, taxis and so on, one has to add the entirely new voice of satellite communication.

All these channels have to be found places in the limited amount of 'radio space' in the electromagnetic wave spectrum, running from the audible range (below 20 kHz) up to the infra-red (above 300 GHz), between limits imposed by technical factors and by the absorption of signals by the atmosphere. The division is made into three main bands. First, the low and medium frequency bands, extending from 10 kHz to 3 MHz. In the early days of radio, transmitters and receivers had only crude tuning circuits and used very low frequencies, blanketing large sections of the spectrum. The advantage was that these waves could travel long distances—some right round the world—reflected by the ionosphere, the electrically-conducting layer of gas in the upper atmosphere. As they were required for communicating with ships at sea, it was all to the good. Because they are the sole kind of radio wave to penetrate water to any extent, they are still employed to signal to submarines. Only eight ground stations would be needed to provide world-wide coverage for a very low frequency navigation system. The military possibilities are obvious (see page 181).

The disadvantages of these waves is that the equipment requires copious amounts of power and large antennae hundreds of metres long. As more radio stations were built mutual interference became more annoying because frequencies overlapped, and it led to the allocation of relatively narrow bands to each station and receiver, and a movement towards higher frequencies.

The medium frequencies, which were brought into play rather later, generally offer good reception up to about 100 km by day and night, and fairly good reception at night up to several hundred km. However, the higher the frequency of a wave the shorter its range, so that at the upper end of this band—around 3 MHz—the range is only a little over the horizon.

High-frequency signals, 3 MHz to 30 MHz, were first turned over to amateur radio enthusiasts because professional engineers did not expect them to travel long distances. In theory they should not. But in fact they do travel well by bouncing off the ionosphere, and they are convenient for communications because they require less transmitting power to cover a given distance, and the antennae needed are only a metre or so long.

There are snags even here, however. High-frequency waves above 30 MHz are not reflected; they pass straight through and are lost in space. Moreover, transmission is disturbed as the ionosphere undulates skittishly on daily, annual, and eleven-year sunspot cycles. Therefore the bands above this frequency, the VHF, UHF, SHF, and the others, are usually reserved for short distances when their stability is an attractive feature.

The search for higher frequencies in the unexplored regions is leading now into the province of microwaves, lasers and optical light guides. These offer much wider communication channels (bandwidths). A single microwave signal, for instance, can carry several hundred television programmes at once, if necessary, or thousands of telephone channels. The laser beam (see page 54), operating at higher frequencies, possesses a yet more ample bandwidth. However, microwaves cannot travel much over the horizon either, and they have to be transmitted from one tower to another so that one aerial can broadcast as far as possible. In Britain, the microwave network links main stations in urban centres, such as the Post Office tower in London, through repeater stations every 40 km or so. The London tower could carry up to 150,000 telephone conversations or 100 television programmes at once, but usually its time is divided between 40 television channels and 100,000 telephone circuits.

As the spectrum becomes more crowded, the narrower each station's band has to be, and the more accurate the receiver has to be to pick out the right message from all the others. For example, over the years the allowed channel width for mobile radio-telephone use has been progressively narrowed from 200 kHz in 1947 to 100, then 50, then 25, and then to 12·5 kHz. Transmitters have to be made so accurate that they can keep within their allotted band, which may mean maintaining their frequency to a few parts in 100,000. So crowded is the short-wave band especially that special techniques are needed to hold frequency to within a few parts in a million.

Technology obviously plays a large part in this exacting field. Formerly, the frequency of a transmitter was controlled by oscillator circuits of coils, wire, and capacitors; now it is usually determined by a quartz crystal, kept at constant temperature, that vibrates at a frequency governed by its mass and size. Then this master frequency is divided and converted to the frequencies required by electronic circuits.

Communication by cable has similar problems. Technology has already overcome the basic difficulties of getting good performance out of long-distance undersea telephone cables, although difficulties of making repeaters (amplifiers fitted at intervals in the cable) strong enough to withstand the vast pressures of the ocean deeps, and reliable to work without failure for decades, were among those that delayed their introduction. Thus the first transatlantic telephone cable was not put into operation until 1956. The problem today is to send as many messages as possible over those cables. But as frequencies can be so much more tightly controlled, and electronic techniques allow time to be measured so accurately, the communications engineer can perform some neat manoeuvres to get more traffic through a given system. Thanks to this sort of technical advance, the cost of using one submarine telephone channel is now about one-twentieth of the cost ten years ago because each cable can carry more channels.

There are commonly two ways of approaching the problem, for transmission over the air, or through a carrier, such as a cable. One is to separate messages by frequency, the other by time. In its simplest form the first involves transmitting many different conversations along the same wire, with each conversation being electrically transposed to a different frequency. At the other end of the wire they are separated by filters and amplifiers and put back to normal-sounding speech. The capacity of this kind of system has been stepped up by the introduction of coaxial cables that can transmit a greater range of frequencies than the ordinary sheathed wire type.

The second method is to allot each message a different time slot in a combined message. The messages are sliced up into tiny bits several thousand times a second, and the bits are slotted in sequence into the transmitted signal. The messages are reconstructed at the far end. It relies on very accurate synchronization of pulses at both ends of the communication channel, only made possible by electronic timing circuits. The latest version of this technique, known as pulse code modulation, is now being introduced on an increasing scale throughout the world.

However signals are transmitted, they must be routed between source and destination by numerous electro-mechanical or other types of switches, except in the case of a broadcast signal which can be picked up by large numbers of people. At present the crucial switching function is often performed by step-by-step electro-mechanical units, redolent of the earliest days of the telephone. These units are relatively slow and noisy, wear out relatively quickly, and are prone to interference from damp and dust. A second generation of switches—the crossbar type—improves on their performance. But as communications systems come under heavier load, and as planners look forward to the high-speed networks and new services of the 1980s, the clearer it becomes that something better than electro-mechanical switches will be needed. Laboratories around the world are devising electronic switching and control equipment (using high-speed circuits of the kind described on page 50) for this purpose (Plate 7.1). The electronic exchange, essentially a computer-controlled switching centre, is far faster in operation, and more flexible.

Electronic exchanges benefit computer systems as well as people. With the growing use of remote computer terminals has come a demand for networks capable of passing messages between them at the high speeds associated with computer processes. In normal speech we convey information at a rate of around 50–100 bits a second (the bit is the unit of information). Computers and other machines can transmit and receive at much higher speeds, however, and so work is now going on into the design and installation of information networks—including electronic exchanges—capable of passing information at rates of hundreds of millions of bits a second.

Microwave radio, and now communications satellites, offer alternative routes for communication. Three satellites hovering about 36,000 km (22,300 miles) out in space could in theory provide radio coverage for the whole world. It is, nonetheless, very costly to put them into this expansive orbit, and in practice more satellites, closer to the earth, are employed (Diagram 7.2).

With the number of telephones in the world doubling every ten or

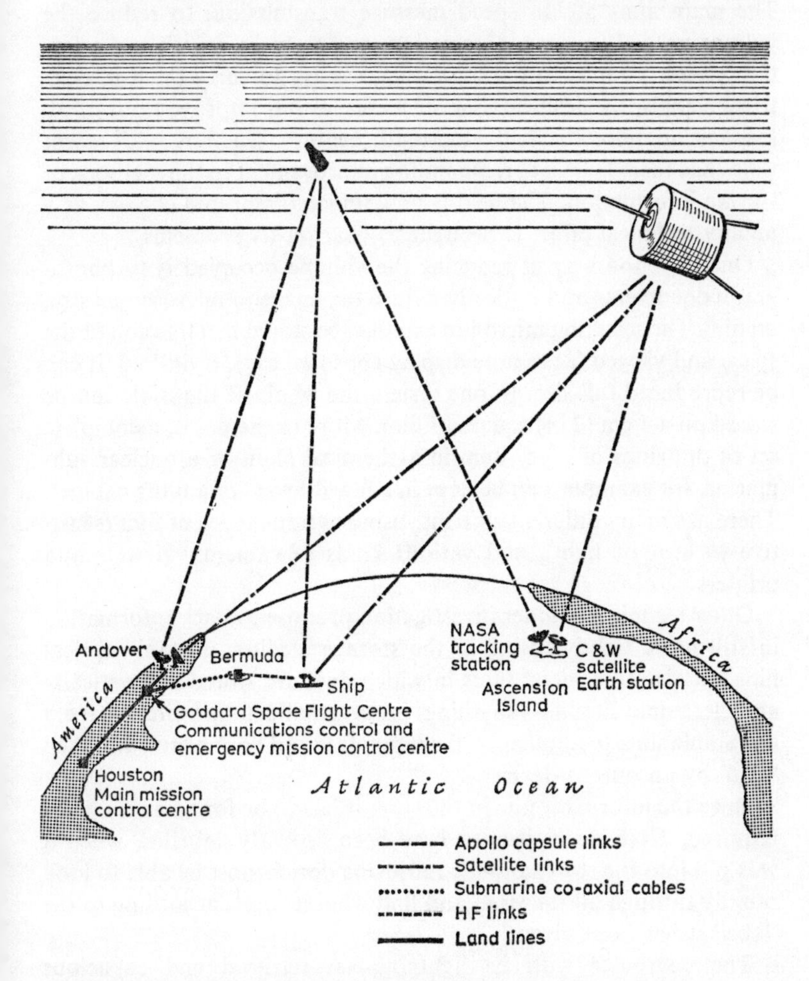

Diagram 7.2 Lines of communication during a U.S. Apollo space mission between U.S. mission control, a communications satellite, ground stations and the spacecraft

eleven years, and long-distance communications traffic quadrupling in about the same time, no technical improvement can come too soon.

Handling Information

What can technology do to help handle the flood of information? The main aims are to speed message transmission, to reduce the volume needed to store information, and to make search and selection easier. Thanks to the discovery of ways to carry them in electrical signals, the transmission of messages has improved out of all recognition since the mid nineteenth century. But not until recent times has there been a corresponding improvement in the methods of storing information. The field is now undergoing rapid change, as a mighty technical effort is brought to bear on its problems.

One common way of reducing the volume occupied is to photograph documents and optically reduce the image to microscopic size, on film. The resulting microfilm can then be stored in a fraction of the space and viewed (at remote display consoles, say). If desired, it can be reproduced full size. In one system the whole of the Bible can be stored on a 5 cm (2 in) square of film. More to the point, a complete set of drawings of every item in a chemical plant or a nuclear submarine, for example, can be kept in a few drawers of a filing cabinet. There are many different systems, using various types of film (sensitive to heat or light), and various kinds of cameras, viewers and printers.

Other techniques under investigation promise to pack information in still more tightly, using as the storage medium not only optical film but also thin metal films in which data are stored magnetically and electronic circuits—anything, in fact, that can be switched from one stable state to another by the input data and back again, and then 'read' by an output device.

Once the information is in the store it has to be found again when required. That means it must have been properly 'labelled' when it was put into the store, and the retrieving device must be able to look rapidly through all the store and find what it wants according to the 'label' it has been given.

The computer with its lightning-fast retrieval and capacious memory devices (see page 93) is an ideal device for storing and retrieving information. Many computer-based reference centres are being set up. While one compiles worldwide information on medicine, say, another collects information on nuclear physics. One pioneer is the National Library of Medicine in the United States, whose computer-based system Medlars handles descriptions of about 250,000 articles a year. The computer sorts these descriptions to produce the text of a secondary journal, *Index Medicus*, and indexes to it. Cumulative issues are generated by the computer with no manual retyping. Medical research workers can get short lists of articles on a particular subject on request. Collaborating agencies in the United States, Britain, Japan and Sweden have either helped prepare information for the system or received copies of the magnetic tapes on which processed information is stored.

The information first has to be prepared for the computer and then entered into its data stores. The 'labelling' task demands considerable human skill, especially where it is a matter of reading through documents such as research reports and indexing them under terms that accurately describe the contents. The computer will only have these terms to go on (they are often called keywords), and if the indexer picks an inapposite keyword the computer will produce a series of inapposite references. It will be many years before indexing can be handed over to computers with any hope of reasonable accuracy. Once the proper information is safely stored in the computer, inquirers can make use of it in two main ways. With the 'current awareness' service, if someone wants to see the latest papers on a certain field of metallurgy as they appear, he can put into the computer a specification of the subjects he is interested in. The computer will then regularly print out the items of interest to him from all those that are fed into its memory on all aspects of metallurgy. The service can be provided for many scientists—or doctors, lawyers, engineers, or anyone else—all at the same time. The computer can even be programmed to address the sheets of paper to each person while printing out the results.

Then there is the 'retrospective search', where the inquirer wants to know what has been published on a certain subject in the last year, or the last five years or ten years, or perhaps as far back as records go. The computer's ability to find such references swiftly is increasingly valuable as the store of knowledge builds up, and as the danger of duplicating work through ignorance grows ever more likely, especially in research.

Computer-based information systems are being developed rapidly and in due course most information required by large organizations at least will be transferred directly between computers. This is the efficient way to transmit information, without all the conventional paperwork, and without the errors of passing through human operators along the communication chain. Large information stores, or databanks, are being established to hold a government's stocks of statistics about its population, a credit company's opinion of the creditworthiness of its clients, a hospital's records of its patients, and so forth. These will multiply and, no doubt, interlink, forming parts of the national communications and data-processing networks in industrialized countries (see page 86).

The hurdles in all these systems come at the input and output points, the interfaces between Man's world and the electronic domain. Within the system, information is stored, processed, and conveyed in 'machine language'. Translation between Man's languages and those of the machines is tedious. We have looked at one aspect—programming languages—(see page 90) and there are many others. However, engineers are continually devising better ways of communicating with the machine. Today's data-processing systems can draw pictures on display tubes and on microfilm, can read printed and some handwritten documents, using light-sensitive circuits (see page 89), and can understand a man typing at his own speed on a keyboard. They cannot yet understand spoken commands, though it is a field of frontier research (see page 92). It may not be many years before a man can hold a conversation with a computer, or operate a process plant by telling it what to do over the telephone. If machines can be made to recognize the differences between the speech of different people, it would provide another method of identification, using a 'voiceprint' in the same way that a fingerprint is now used.

The Television Set and Beyond

The television set like the car is a symbol of twentieth-century technology. But unlike the car, its influence is just beginning to become apparent. We are only starting to appreciate its unique qualities and to realize how it could be used in hitherto untapped ways. Yet it is the most powerful of the existing communications media and—until someone invents a medium that can transmit sensation directly—the most effective. Its great feature is that it can bring both sight and sound from the outside into the home, allowing a man to range the world from his armchair. It is a source of entertainment and news that all can share from the toddler to his grandparents. At the turn of a switch they can be in the front row of the theatre, in the Wild West, or even on the moon. No other medium can perform all these miracles in the home.

The introduction of colour brought in another attraction, in the early 1950s in the United States, later in Japan, and later still in Europe. Provided there is international agreement on broadcasting standards, or devices can be found to convert from one set of standards to another, colour television will be a universal medium of great popularity. There is more to come, however. With the aid of the new video-recording systems being introduced now, the ordinary person can buy prerecorded programmes on film to run on his own set, and record and replay his favourite television programmes in colour or black and white. Video tape cassettes (about the size of a paperback book) and video records are expensive, as are the auxiliary playing units, and initially the cost is likely to prevent all but institutions like schools, universities and colleges, and companies from adopting them, but mass production could lower the cost appreciably. One type of system, that made by RCA of the United States, uses as the medium a cheap vinyl film embossed with a hologram of the original picture and a laser to reconstruct the picture in the home viewer—the first domestic use of the laser and holography. A two-way television channel would enable a viewer to select a programme from a central library, and then watch it played on his television set, without stirring from his house.

Another ancillary is a copier unit that enables a television set to produce 'printed' material. This allows a newspaper office, say, to send out a paper page by page to the television-copier, which then prints them right in the home. They could be transmitted while the programmes were off the air, or while programmes were actually being received, the information being 'slotted' into 'blanks' in the television signal. The set could in this way become a fount of printed news and comment, and so a direct competitor to newspapers and magazines. Until now, these have scored in that the reader can scan, mull over and return to them at his own pace, and then file them. If visual images in television have great instantaneous impact, they drag the viewer along at the camera's pace and leave him nothing to return to later. A television-copier could prepare a family's newspaper several times a day, untrammelled by all the clumsy distribution chain that is needed to get papers and magazines to their readers. It is a challenge that printing and publishing—for all its innovations in the shape of computer type-setting and photographic processing—may have to confront before long.

Impressive though they are, these new features built around the television set are no more than forerunners of a host of other services that technology will be able to bring into the home. In communications, promise is almost as real as reality, so quickly does technology force discovery into mass production. With the arrival of cheap electronic devices in plentiful quantities, communications will be further improved in many ways.

Research has already produced the video-telephone, equipped with a small television camera and a cathode-ray-tube screen, so that people using the instrument can see each other while they talk. This sort of device, connected to a communications network of the new kind, offers almost endless scope for forward-looking thinkers. For example, a video-telephone equipped with a pushbutton keyboard would allow the housewife to view goods in the local supermarket and place her order without leaving the house, authorizing her payments with a 'magnetic money' or credit card. The bill would be sent to a local computer, which would arrange for payment without any cash having to change hands. Fuel meters could be read centrally by telemetering attachments on the video-telephone. By keying the number of the local library, a schoolboy could work through a pre-recorded lesson in mathematics for homework; the businessman could dial another number and see the latest stock market prices; the doctor could consult specialist opinion on a case without leaving his surgery.

In the future the cathode-ray display may be replaced by a solid-state display in colour (see page 53). It will no doubt be possible, if expensive, for the home to have a colour television right across one wall; such displays are already made in small numbers for business meetings, schools and television studios. For businesses, there will be networks of conference rooms in major cities across a continent, each in effect a studio fitted with a special projector that casts a giant colour picture on to a cinema-like screen. With all linked into one huge 'conference', there would be no need for each man to waste time travelling to a central point to meet everyone else. He would just travel the shorter distance to his local conference room. The television links would do the rest. Even outside the room he could keep in touch with headquarters through a cordless wristwatch-telephone only a few centimetres square.

In the not too distant future we may be able to live where we like instead of having to find a home within easy travelling distance of our jobs. We could communicate with offices, clients or customers from home, travelling merely for pleasure. If this kind of 'decentralization' were widely adopted—as it was in the early days of the Industrial Revolution, with its cottage workers—it would help to reduce the chronic congestion of many city centres. Whether the people concerned would want to work in that kind of isolation is another matter.

Using Information

Modern communications has made us much better informed about the world than any previous age. Books, radio, and television reach all but the most isolated communities. In the deepest jungles people have been introduced to the wonders of the transistor radio. The telephone is not far behind. Gramophone records circulate as a kind of global currency, carrying the imprint of the world's top entertainers. Technology, through all these media and others, removes the old constraints to communication that hindered the growth of earlier communities. It multiplies the contacts between people by increasing both the distance over which they can communicate (e.g. the telephone) and the number of people with whom they can be in simultaneous contact (e.g. broadcasting).

Being an inquisitive creature, Man wants to know what is happening around him. Being a gregarious one, he wants to keep in touch with others of his kind. It was a great step forward in his development when he learned to speak, instead of communicating in grunt and gibber. Yet he could not communicate in writing. It was another great step when around 3000 B.C. the Sumerians, living in the valleys of the Tigris and Euphrates, invented writing, and so became able to record their experiences in communicable form. Before that, prehistory stretches in dark aeons of ignorance, illuminated only in flashes by pictures that are spirited but beyond true comprehension.

Using only natural means of expression—speech, posture, and

movement, chiefly—a person's range of communication is severely limited to those that can see and hear him at a time. A soldier's parade-ground bellow carries no further than a few hundred metres. Technology has widened the circle dramatically. The invention of the loudspeaker, for instance, made it possible for an audience of 500,000 at a pop festival to be in rapport with the performer. Broadcasting, though it cannot preserve the same atmosphere of a 'live' performance, enlarges the audience by another order of magnitude. When the United States landed men on the moon in 1969, there were estimated to be 600 million people—a fifth of the human race—watching on their television screens (there might have been more had not China's millions been kept in ignorance). Almost as many watched the opening of the football World Cup in Mexico City in 1970.

Now technology has so improved communications that we are able to keep up to date with events on the other side of the globe, and even in the solar system and beyond. Everyone, it seems, wishes to communicate with others, to see and hear what they are doing.

But there is one major difference between these technical modes of communication and our natural methods. Technical modes are at present essentially one-way. Information flows out from the broadcasting station to the listening or watching public. Participation by the multitude is as yet impossible. People are kept in touch, and yet feel isolated because they cannot respond. The individual is isolated from the source and amid a constant stream of material—printed as well as broadcast—feels he cannot alter its style or content. An ineffectual, passive spectator, he is only valuable as another digit in the massed ranks of the audience figures.

There are other drawbacks. A communication centre with a mass audience wields great power. The material it chooses to disseminate will mould attitudes and opinions; it may be critical of government; it may laud one group of people louder than others; it may allow commercial companies to boast of the attractions of their products. Whatever it is, technical devices such as the radio and the television bring a steady stream of entertainment, education, etc., into the home's private environment. Those who control them have unique opportunities for pursuing their own ends. Of course, this may have no sinister overtones. There are ample opportunities for putting out educational programmes, a boon to older people who could not enter college or other training when they were young.

Informed reporting provides a means for the electorate in a democratic country to keep a check on the actions and policies of their leaders. The broadcasting of debates in provincial or national government, and of searching interviews with political leaders, is an unrivalled way of subjecting them to public scrutiny. Nevertheless, the dangers do exist.

On the personal scale, people who can be amused by the world's top performers at the turn of a switch lose the wish to do things themselves. Evenings around the piano or the card table, and hobbies, recreations and games are steadily losing ground among large sections of the population to the radio, cinema and television. As a child grows up, in one of the industrialized countries at least, television personalities can become as real to him as his family. He can certainly see as much of them. In the United States, a home television set is watched for an estimated average of 5½ hours a day. That means by the time a child enters school he will have spent 3000 to 4000 hours in front of the television set. By the time he leaves high school he will probably have seen 15,000 hours of television, compared with the 10,000 hours of formal education in school. Very likely, he will come to adopt the values and standards of the television world as his own. He will come to accept the television's view of life (especially where events are outside his direct experience) as the true one, and unless he is careful he will find the fleeting figures of the screen more vital and interesting than the real ones around him. Family ties are all too easily weakened in a household ruled by television.

Mass communication may also have quite unexpected effects. We can hardly blame a man in a developing country for demanding steaks to supplement his meagre diet when he has watched films from the rich world where a steak is an everyday dish. The same goes for a whole variety of 'sophisticated' products. The claim that steaks and other goods are not available is unlikely to satisfy him. Communications, by showing poor people what affluence is, has probably done more to breed social unrest than any other factor. Yet, on the other hand, television programmes relayed by satellite could provide the cheapest and most efficient way of teaching about agriculture, the three Rs, and health matters in remote villages.

So, like technology overall, communications devices are basically neutral. They will do what their masters tell them to do. There is one thing about them that is not in doubt, however. They will assume a major role in world affairs as technology becomes more reliant on information. A government, supplied with immediate statistics on trade, currency flows, births and deaths, housing construction and the rest in its own country and others could shape policy responsively to suit real conditions. A multinational corporation, similarly fed with up-to-date information on raw materials prices, market conditions, shipping rates and so on across the world, can throw its corporate energies into the most profitable directions. In the office, business machines have removed much of the drudgery. A wide variety of calculating machines can perform tasks ranging from complex accounts analyses to simple arithmetic. Techniques of wet and dry duplicating, notably xerography, have transformed the task of reproducing documents. Nothing has so far replaced the typist, though there is no reason why work into machine recognition of sounds should not lead to an 'audiotyper' that would convert clearly articulated speech into text. Machines also assist with information retrieval. In a typical office, staff spend 20 to 30 per cent of their working time simply looking for information and anything that reduces this unproductive time obviously helps the work flow.

In school, university or college closed-circuit television raises the lecturer's productivity as he can talk to many more students at once. Indeed, many colleges can co-operate in showing a lecture pre-recorded by the best lecturer of all the combined staff (or even by an outside expert), leaving the other staff in their several lecture rooms to go round and give students individual tuition.

Just as the steam engine, earthmoving equipment and machine tools raised the productivity of manual workers during the first Industrial Revolution enough for certain industrialized societies to buy good things from all over the world, so data processing machines—computers and communications systems—of the current information revolution are increasing the productivity of those working with information, whether they are performing invoice calculations, working out stresses in designing buildings, cars and aircraft or booking aircraft seats.

The main barriers in the way of developing free interchange and application of information still lie in the interface area (see page 160), and in the area of technical standards. It has proved extremely difficult to get different countries to agree on common practices and standards. At the moment, an inventor still has to take out patents in different countries to protect an invention, and even then some may not respect the claim. Printed material and films are covered by copyright in certain countries, but again not all respect it. Lawyers have not been able to settle the knotty problem of whether computer software, punched tape and magnetic tape should be protected or not, and the means by which this might be done. In spite of a growing trade, no one has yet discovered how to arrange a lively interchange and efficient processing of information while keeping some of it confidential for military or commercial reasons.

There are many imperfections in Man's use of information. One is especially serious, and it concerns the fragmentation of knowledge. The more Man knows about the environment, the better able he is to control it. However, as the sum total of knowledge increases, each individual knows less and less of the whole—that is, he is ignorant of most of it. As education and work are made more specialized, ignorance will become more prevalent. Although the specialist can apply technology precisely to a specific problem, he lacks the awareness to foresee its ramifications. People who can visualize the overall impact of technological change on social, environmental and economic factors are few and far between. Information systems may help present the relevant data on which to base a decision, but ultimately the decision on how to use technology with its great flood of information is ours.

8 · Technology and Transport

The Lure of Speed

Man is restless, inquisitive, forever moving around the globe—outside it as well—looking for fresh realms to conquer. His deep-seated urge to prospect the environment is surely at the heart of technology. He has travelled over the oceans for the past 8000 years; on wheeled land vehicles for the past 5000; under the sea and through the air for 200, and into space for 10. He has used the new-found impetus of the internal-combustion engine to propel himself through the air, over land and sea, and then the augmented impulse of the rocket motor to launch himself into space.

The way in which technology has shrunk the globe in a few centuries is astounding. Five hundred years ago it took Marco Polo and his companions three years to travel from Europe to China and back. Now a supersonic aircraft can fly from Europe to Australia in twelve hours, and a manned spacecraft can circle the globe in about eighty minutes. High-speed craft whisk ordinary folk about at speeds that were solely for the pioneer record-breakers one or two generations ago.

In all this exploration, speed has exerted a strange fascination, luring men on to ever greater exertions. The aerospace industry (as it has become) has done most to push back the barriers to rapid travel. Only in 1903—not long in terms of human progress—did Orville and Wilbur Wright first fly their rickety, powered contraption at Kitty Hawk in North Carolina. It flew at not much more than 16 km per hour (10 miles an hour) and soon came to earth again. A whole new industry has since sprung up around the building and flying of such craft; a new form of warfare has been evolved with manned flying machines; and fresh, glamorous careers for such people as pilots and air hostesses have been created to serve them.

Speeds have steadily risen. In the 1920s airliners cruised at around 160 km per hour (100 mph). By the mid 1940s they were approaching 480 km per hour (300 mph). Today's jets cruise routinely at nearer 970 km per hour (600 mph). The supersonic transports, such as Concorde, have raised that to over 2250 km per hour (1400 mph). Military planes and development craft fly faster still, as we know.

The American experimental X-15 series of 'piloted missiles' rocket aircraft reached speeds in excess of 6500 km per hour (5000 mph). A second generation supersonic airliner based on current research could cruise at four times the speed of sound (Mach 4) at around 30,000 m (100,000 ft). Before the end of the century the hypersonic airliner, cruising at Mach 12 or nearly 13,000 km per hour at 43,000 m (8000 mph at 140,000 ft), could bring Sydney to within two hours of London. Intercontinental ballistic missiles (ICBMs) can attain similar speeds in their trajectory through the atmosphere. In space, where there is no air resistance, craft have reached 39,000 km per hour (24,000 mph) and more.

In this evolution technology has played a major part. It can be seen in the adoption of new materials, notably aluminium and magnesium alloys, for the construction of the airframe and engine. Composites (see page 76) made from thin fibres in a more pliant matrix enable aerospace designers and engineers to make light, stiff, strong structures that can be moulded or perhaps woven into the right shapes. It can be seen in the development of the stressed-skin monoplane structure with swept wings and retractable undercarriage. Just as important is the influence of technology on powerplants; and above all the invention of the jet engine (gas turbine) in Britain by Sir Frank Whittle in the 1920s and 1930s, and the rocket, which was used by Germany during the Second World War in its hideous flying bombs, forerunners of the ICBMs. All these enable aircraft and spacecraft to fly faster, further and with greater payloads.

In commercial use, the jet engine, by enabling airliners to cruise close to 970 km per hour (600 mph), has cut air-journey times dramatically, bringing air travel within the reach of far more ordinary people for a reasonable cost. On the transatlantic route, for instance, flight times have been almost halved from sixteen hours less than fifty years ago to eight hours. Essentially, speed is a matter of more potent engines: the greater the thrust, the faster the aircraft. Advances in materials and other fields allow designers to extract better performance from engines so that tomorrow's models will be smaller and lighter for a given thrust, running hotter than today's. Development is still proceeding with the jet engine variant, the turbofan or bypass turbojet, in which much of the air, propelled through a cowling round the outside of the engine by a large fan, mixes with the hot exhaust gas. This offers a better takeoff thrust and lower fuel consumption than a 'straight' jet. For hypersonic flight the aircraft would have to be powered by a ramjet, as the ordinary gas turbine cannot fly at much more than Mach 4. But a practical ceiling is likely to be reached with Mach 10, which aircraft will hardly be able to attain in a journey taking less than an hour between London and New York.

The money and manpower devoted to this one end—increasing speed in the air—is immense. The stimulus has come mainly from military needs during two world wars. Civil impetus has developed afterwards. The main reason for the much increased cost of today's aircraft is largely due to their greater complexity. Present-day aeroplanes like the Lightning fighter are intricate, interrelated masses of electrical and electronic, hydraulic and mechanical equipment. Whereas the production cost of a World War II Spitfire was about £5000, a Lightning runs to well over £500,000. The development cost of an airliner now runs into tens or even hundreds of millions of pounds, making state aid—and management involvement—necessary.

There is no denying that aerospace technology has done what it set out to do—essentially, to raise speed of travel—superbly well. If a Mach 10 airliner did go into service before the end of the century (although this is doubtful in view of current public disenchantment with supersonic aircraft), it would enable Man to travel at a speed approaching 1000 times faster than he could ever have done before the introduction of mechanical vehicles early in the nineteenth century.

Besides increases in speed, the size and carrying capacity of aircraft have grown rapidly. The introduction of the mammoth 'jumbo jet' aircraft in 1970 marked the latest and most dramatic change, and the real start of mass transport in the air. The first, the Boeing 747, came into service early that year. Its dimensions are staggering. Tall as a five-storey building, it is longer than the Wright brothers' entire flight, only sixty-five years before. Carrying 350 to 490 passengers on two decks it cruises at 1000 km per hour (625 mph) for upwards of 8000 km (5000 miles) without refuelling. Each of its Pratt and Whitney engines is three times more powerful than conventional turbofan engines. Currently, the Boeing is being joined by other jumbos—at least two American ones, and possibly others. These jumbo jets should prove more efficient than the current generation of airliners, partly because of better engines, partly from using tougher, lighter materials, and partly through better aeronautical design, using the latest lift-enhancing devices such as leading-edge slats and trailing-edge flaps.

The military stimulus still exists. The world's largest plane at present, the American Lockheed C-5A Galaxy jet transport, can carry 100 tons of cargo or 800 passengers (Plate 8.1). It was built for the U.S. Air Force, but in a civil role it could doubtless carry almost as much, depending on the standards of comfort called for. The trend towards larger aircraft will probably continue. Aircraft with a thousand seats are said to be technically feasible, and would be cheaper to build than today's craft in similar numbers. Operating costs of large aircraft fall with increasing weight up to around 0·7 million kg (1·6 million lb), twice as heavy as the Galaxy, and although the improvement at the heaviest weights is not large, it is significant.

Behind the impressive statistics lurk many problems. On the ground, already crowded airports have to cope with the rush of nearly 400 passengers disembarking within the space of a few minutes, with their luggage. Massive new terminals capable of handling many jumbos at once have had to be built at immense cost to cater for the influx. World airline traffic is rising fast. According to the U.N.'s International Civil Aviation Organization, world air traffic in 1968 totalled 261 million passengers, a 12 per cent increase on 1967 (excluding the Soviet Union). The world total (excluding China but including the Soviet Union) was probably 320 million. Over the next ten years or so scheduled passenger traffic is expected to continue to grow at over 10 per cent a year—doubling in about five years—while freight traffic will grow still faster, doubling in about four years.

While airports may be hard-pressed to cope with all this traffic on the ground, it is far worse in the air. Already, after only sixty years of flying, the air around major airports in particular is chronically overcrowded. Planes have to queue to take off and land, circling round in the air in 'stacks' as they wait. The United States and central Europe are especially afflicted. The capacity at John F. Kennedy International Airport, New York, is 260,000 aircraft operations a year, if the planes are to be kept on schedule: yet in 1968 and 1969 the airport was handling 500,000. Obviously there are delays. Already five American airports are said to work at saturation during peak hours: unless there is major expansion, there will be twenty by 1980.

The situation is more harrowing for airport staff trying to control planes without undue delays and without compromising safety. As air traffic builds up, collisions become more likely. Up to the time of writing, the worst disaster in airline history was said to be that in October 1972 when 176 people were killed in the Soviet Union, when an Ilyushin-62 jetliner coming from Leningrad crashed while trying to land in bad weather at Moscow's international airport. It made three approaches in the heavy rain, then crashed on the fourth attempt. An accident involving a jumbo could be far worse. An accident involving two will become more probable as their numbers increase: it is particularly likely to occur around congested airports which are generally surrounded by densely populated areas. Casualties could total more than 600. Even if there were no serious injuries, the financial losses are potentially tremendous. The insurance cover on a Boeing 747 amounts to perhaps as much as £50 million—including the hull, full cover on a complement of around 360 passengers and crew, and the damage that might be caused on the ground. Insurance agents are anxious about having to insure such colossal sums, not least in the highjacking era, while governments and business organizations are having to arrange that they do not send large parties of their senior personnel on a single plane to avoid the risk of losing them all in a catastrophic accident.

The problems of congestion would be eased if there were more airports, or greater space for existing airports. The size and weight of conventional airliners now in service, and the noise of their engines, forces airports to become larger. New ones have to seek sites further away from the cities they are meant to serve. Already someone flying between, say, central London and central Paris spends less than a quarter of the average total time of about five hours actually in the air. The rest is taken up passing through the airport formalities and travelling between airport and city centre. A large jet airliner weighing around 145,150 kg (320,000 lb) on take-off needs about 2700 m (9000 ft) of runway to work up to its take-off speed of about 220–300 km per hour (145–225 mph). At international airports the standard length of the main runway is 3050 m or so (10,000 ft). It is no easy thing to find the required land in a place where the residents will not object too volubly. The presence of the airport exacts a substantial penalty on the social and economic balance of its neighbourhood.

Until recently airlines were free to travel where and how they liked. Now they are starting to have to bear the costs of 'sociological factors', including aircraft noise, which involves airlines in costs for the insulation of buildings as well as the settlement of legal actions over noise; this is aggravated by the booms caused by supersonic aircraft. Although there are generally noise regulations in force at airports, and manufacturers are aiming to make new generations of airliners much quieter from the outset, the efficacy of the procedure is in doubt so far as the residents around airports are concerned: the planes themselves may be less noisy, but there will be more of them as air traffic increases. And in any case it will not mitigate the effect of the sonic boom. Though a great deal of effort is going into methods of reducing boom intensity, nothing has yet appeared that looks practicable. The other major sociological factor is the need to ease pollution from jet engines by introducing smoke-reduction and other devices (see page 268).

All these modifications to the aircraft to make them less of a social nuisance have to be paid for. It is up to society to say how the 'social costs' are allocated between the manufacturers, the airlines and the public.

The Turmoil of the City

Such has been the preoccupation with the heady new aerospace technologies that more attention has been given to adding a few kilometres an hour to an aeroplane's speed—a craft in which only a minute fraction of the world's population can hope to travel—than in attending to the weighty problems of the prosaic forms of transport on which most people depend in their everyday life. While aeroplanes gracefully whine overhead, the world's cities are strangled by growing hordes of cars, lorries, trams, bicycles, buses, vans, rickshaws—floods of traffic that they were never meant to take.

The problem is not a new one. In the first century A.D., goods vehicles were banned from the streets of Rome during daylight. Obviously, though, it has got far worse since the internal-combustion engine replaced the horse as the power unit for vehicles and Henry Ford started the application of mass production in the motor industry in the early 1900s. Yet, until the past decade, the modern world has turned a blind eye on the problems, largely because of its worship of the motor car.

The car represents everything that most people want from a mode of transport. In the mid nineteenth century the railways fulfilled the same sort of role. For most people, the ability to travel at great speed over long distances (as the aircraft does) is not the point. What they want is a cheap, reliable and above all convenient means of door-to-door transport, giving them freedom of movement. This the car does, amply. The fact that it also acts as a sanctuary of a kind—a form of mobile private environment that shuts off the individual's surroundings from the rest of the community—is another point in its favour. In addition, it gives people an outlet for repressed aggression, sex urges and other motive forces. The car (together with the television set) is one of the great liberating agents that modern technology has provided for the ordinary man.

Small wonder, then, that the car industry, together with its ancillary trades, is a pillar of modern technological society.

To give the car manufacturers credit, they have continually complied with people's wishes by introducing engines with better performance and lower fuel consumption, better lights and brakes, better visibility, and so on; cars have become faster, more powerful, roomier. Occasionally, they have hit on a real innovation, such as the Mini, with its transverse engine and front-wheel drive, a nippy city car if ever there was one (Plate 8.2). Then there is the adoption of the rotary engine (see page 142). But in general it is a question of evolution not revolution, as with the application of integrated circuits for various tasks under the bonnet (see page 49). To generalize, the car industry—and hence the car itself—has not been responsive to the changing conditions in which it now operates. In particular, the industry has hardly begun to tackle the problems of safety and air pollution, and the initiative has had to come from outside. Not till the late 1960s did the United States (surely the most car-conscious nation in the world) introduce safety legislation for cars. The requirements for 1969 and 1970 models included interior padded surfaces, steering wheels that collapse under impact, permanent head rests for driver and front-seat passenger to prevent neck injuries caused by something running into the back of the car, and windscreens that can retain 75 per cent of their area after a collision. Such new features should protect people in spite of themselves, which has not been true of previous safety items: although seat belts are compulsory in the United States, it is thought that 80 per cent of drivers ignore them.

There is mounting pressure on the industry for the control of vehicle exhaust pollution, too, and it is forcing car makers and petrol refiners to change their ways. The first anti-pollution moves came in the early 1960s, since when regulations have stiffened considerably. The US Government has decreed, for instance, that the exhaust emissions from cars built after 1st January, 1975 shall be reduced by 90 per cent of the 1968 level. It means that cars will have to be fitted with catalytic converters and other devices in their exhaust systems to ensure that combustion of the fuel takes place more completely. Extensive design changes will have to be made to engines, involving lower compression ratios and hence less power—a reversal of the technological trend. Oil companies have agreed to market petrol without the lead formerly added as an 'anti-knock', because the lead poisons catalytic converters after only several thousand kilometres of driving. The changes will be expensive, and someone—probably the motorist—will have to pay for them.

So fierce did the anti-pollution fervour run in the United States once it had taken hold that in the late 1960s motorists started to consider other forms of propulsion for vehicles. One was the steam engine, which had preceded the internal-combustion engine on British roads in vintage motoring days. It burns an inexpensive, low-grade paraffin so completely that it produces very little pollution. The electric car, another candidate, cannot yet match the performance of the car driven by an internal-combustion engine. Electric milk floats and trucks are well-tried vehicles testifying to the worth of electric propulsion, but their batteries have to be recharged frequently, and their speeds are low. So far no battery or fuel cell—another possibility as a self-contained power source—has been produced that is both light and cheap, and capable of supplying the peak currents for swift acceleration, and also has the stamina to go long distances without recharging.

Not that interest has evaporated. Many car manufacturers, Ford and General Motors among them, and makers of batteries and electric traction equipment have invested heavily in development. However, the technologies of electric motors and controls are currently ahead of that in power sources. Much more research is needed, and governments that are anxious to see air pollution reduced could further that aim by investing in research. The goal of reducing air pollution is a social—and not an immediately profitable—one. There has to be some form of inducement to provide the necessary boost. Before too long it is likely that all major countries will have imposed rigorous standards of safety and pollution for vehicles, especially cars.

Worst of all, cars can be made by highly automated industry much more rapidly for the roads available (Diagram 8.1). Overcrowding is worst in the cities where the great majority of people live (see page 248). The turmoil of city transport demonstrates clearly the conflict between mass needs and individual preference. The interests of the community dictate that large volumes of densely-loaded passenger traffic should flow freely in well-defined streams, preferably without crossing the paths of other streams. Logically, therefore, everyone ought to travel by public transport. Yet the motorist likes the freedom that his car gives him from the fixed routes and timetables of the public system. He drives blithely into the city, his car often empty save for himself, and curses all the other cars for causing congestion. It is believed that peak-hour congestion in such cities as London, New York and Tokyo is caused by a minority—say 10 per cent—of all commuters who insist on driving to work whilc trains and buses together carry all the rest in a much smaller space. Buses running on a reserved lane can move 25,000

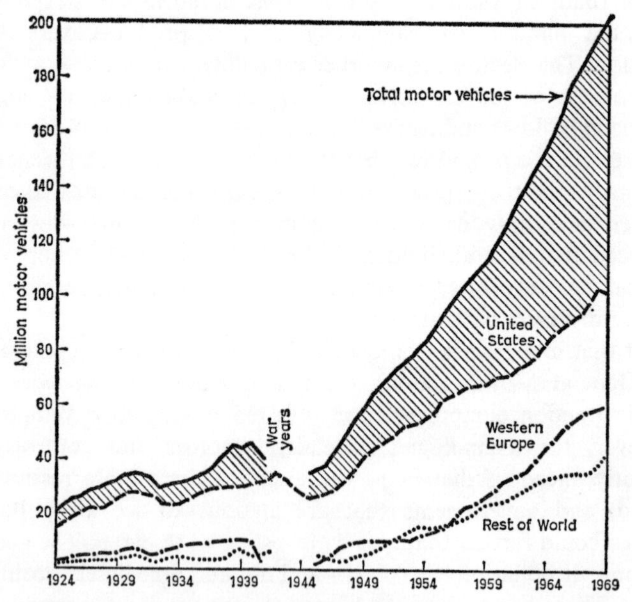

Diagram 8.1 Total registrations of motor vehicles in the free world
Petroleum Information Bureau—Australia

to 30,000 sitting passengers an hour, whereas cars, each carrying the city average of around 1·5 passengers, can only take 3000 people per lane. Usually, however, the motorist prefers to sit it out, not forsaking his car until congestion reaches alarming proportions.

No city has so far reconciled the need to arrange for mass movement with that to give everyone freedom of choice. It is, of course, impossible. With car ownership rising, it seems inevitable that either cities will have to be redesigned around the motor vehicle, or vehicles will have to be banned from the cities. In Britain, there is one car for every six people: in the United States there is one for every three. In the city of Los Angeles, the supreme example of surrender to the motor car, there is almost one for one. Many cities are trying out compromise 'solutions'. Computer control of traffic flow, assisted by closed-circuit television (see Plate 8.3), can and does bring marginal improvements in flow across parts of, or even across whole, cities. There are more radical ideas, though. In the United States, Britain and elsewhere serious thought is being given to 'taxis' and 'buses' that are automatically steered on computer-picked routes. The guided bus could be driven as normal on ordinary roads but could transfer to special tracks, elevated when necessary, where it would be guided electronically by a probe following a central rail. Although it would be slower than certain other forms of transport—like the monorail, which is being tried in a limited number of installations—it does not share the disadvantage of being tied to a fixed track and stations. The taxi might be directed by a computer, programmed by the passengers' tickets, that would be inserted into the taxi's guidance unit.

These are by no means the only ideas. Another, the moving pavement, is a broad, continuously moving band which carries people along, much in the manner of a horizontal escalator. Another type would consist of 'pods' or small cabins holding one, two or more people. Running on a track, perhaps like a monorail, they would also fulfil the role of driverless taxis. One system has small plastic cabins, electrically propelled, that automatically follow a fixed track to the destination selected by the customer on a map of the network. A central computer would ensure that there is a permanent supply of waiting cabins at all stops by sending empty ones as and when required. With an estimated average speed from start to finish of nearly 50 km per hour (30 mph), it compares well with present values of 35 km per hour (22 mph) for underground trains, 20 km per hour

(12 mph) for cars, and 15 km per hour (9 mph) for buses and trams.

One major difficulty is getting people on and off. The moving pavement can have slower feeder pavements alongside onto which a person steps at the start of his journey to accelerate to the speed of the main band. The pod and carriage 'railways' would also have subsidiary acceleration tracks, joining the main track, and there would be ingenious mechanisms for attaching the pod or carriage to the main track, and uncoupling it at the right stop.

Junctions are in fact the bane of transport systems. Eliminate them and segregate different traffic directions, and travel becomes quicker—as motorways prove—and, one hopes, safer too. That is all right in theory. In practice it is a fight for land between open space, buildings, roads and railways. Ground congestion, particularly in industrial countries, taxes the ingenuity of transport planners. One alternative is to build up, as is done with motorways fairly frequently. The other alternative is to build underground; however, this is expensive, and tunnels have to avoid all the other underground services—sewers and waterways, electricity and gas, telephone, etc. Even so, new underground railways are continually being opened around the world to serve city needs.

All these new systems must be fitted into the city's existing shape, and around its existing transport modes. All the current land-transport systems demand massive investment in such structures as roads and railways that become part of the environment, part of the way of life. It is very difficult to move them or to change them radically. They last longer than is technically efficient: but technology has to bow to civic circumstance. Thus the traveller will encounter the odd sights of a monorail whisking passengers from Tokyo airport above the cars on the motorway and rapid express trains forging along the centre strip between motorways in Chicago. The lifetime of roads and buildings is upwards of 50 years. To shorten it significantly would be excessively costly. On the other hand, the lifetime of vehicles is less than 10 years. This implies that—as has already happened with the car—a fundamental change in the type of vehicle in mass use could render obsolete the whole of the environment built for them, for the best part of a century.

Inter-City Travel

While the technological means are ready for the cities but are held up by the massive costs and the social upheaval involved, technical solutions to inter-city travel problems are somewhat easier to apply. There is plenty of technical choice among the different systems. Among the chief ones are aircraft, including helicopters and other forms of vertical-take-off craft; fast railways, including conventional trains with existing types of drive or jet engines, say, to give them an extra turn of speed; and novel tracked vehicles (such as the 'pneumatic tube' train, which is propelled by air pressure differences through a smooth, closely fitting tube) or overhead-rail types, where the carriages hang beneath a rail from a wheeled trolley. Most radical of all perhaps is the tracked hovercraft, which floats on an air cushion (Diagram 8.2). It could be driven by various means, such as the jet engine—making the comparison with an aircraft even more apt—and the linear induction motor.

The vertical-take-off craft could fly from decks in the heart of a

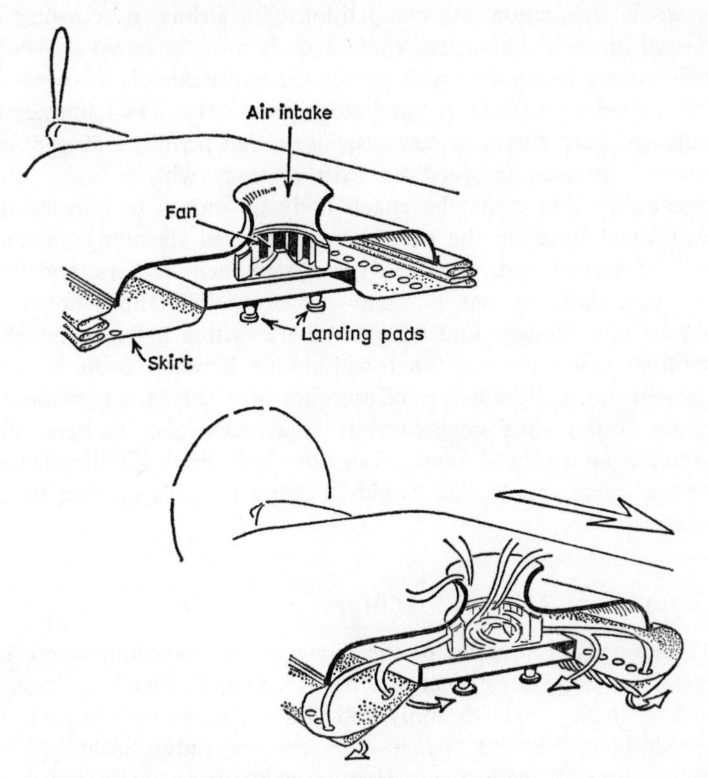

Diagram 8.2 The hovercraft principle of supporting a craft on a cushion of air, which is retained by a flexible skirt

city, maybe on top of skyscraper blocks, to other city centres, saving the time and trouble now expended in travelling between city centre and airport. The technical problems are stiff. Its engines have to yield about three times the thrust of those designed for a conventional airliner because the load has to be lifted vertically without any help from lift created by forward motion. As the craft would have to fly into densely populated areas noise will have to be minimized, and the aircraft will have to be exceedingly reliable.

The hovertrain is a powerful ground-based competitor, one of the new breed of machines making use of the floating, low-friction properties of a cushion of air pumped under the base of the vehicle by powerful fans. Given the right conditions, and in particular a concrete track accurately laid over long distances, it is capable of cruising speeds in excess of 400 km per hour (250 mph) (Plate 8.4). It would thus come into competition with airlines over routes of several hundred kilometres over land. It poses a threat to which railways are responding with redesigned trains capable of speeds of 240 km per hour (150 mph) and more, given either new-laid track to iron out sharp curves or new suspension that permits a 50 per cent increase in average speed on existing track with a body-tilting mechanism that cants the coach body on curves to balance the centrifugal force on the passengers. Improved signalling methods will be needed, too. All these types have their backers, and it is probable that any one of them—or any combination—could be successfully brought into widespread use within a few years of a contract being placed. What will decide between them is cost, convenience and the degree of nuisance they cause the community. Noise is the chief concern, but appearance also matters. For example, an overhead monorail or hovertrain track winding its way across scenic countryside would in many people's opinion be an eyesore.

Guiding the Moving Machine

Traditionally, it was up to the navigator to make out where his vessel was and steer it clear of other vessels and obstacles. The first, and for many years the only, aid was the magnetic compass. As speeds increased and voyages became more adventurous, Man's senses proved inadequate. Human skills had to be aided by machine. When Man took to the skies, moving rapidly in three dimensions, mechanical navigation became of prime importance, especially because there were no tracks to follow and ground landmarks were frequently obscured. Nor are there tracks in the sea for surface vessels to follow; and the submarine must move safely in three dimensions very often without sight of the surface for hours—or days—on end.

The hallowed way of locating position—still adopted even in spacecraft for checking the accuracy of mechanical systems, and in aircraft flying over long stretches of featureless desert or sea—is to take sextant sightings of the sun or stars and combine these with chronometer readings. To supplement that, it is possible to create 'tracks' through air and sea by means of radio beacons that transmit guiding signals. Localities all over the world are equipped with several systems of this kind. Some beacons send out very high frequency signals, of limited range (see page 154), and thus need to be installed every few hundred kilometres or so on land along air routes. Others employ lower frequencies and hence have longer range. The United States Omega system is said to provide from no more than eight beacons around the world an all-weather global navigation service equally suitable for air, surface, and submarine use. The latest 'beacon' is a satellite from which the ship or aircraft can pick up a bearing. The new Cunard liner *Queen Elizabeth 2* is the first passenger vessel in the world capable of navigating by satellite, fixing its position to within 0·16 km (1/10th of a mile) in all kinds of weather anywhere on the high seas. The system uses data from polar-orbiting satellites of the U.S. Navy's navigation system. Tracked from the ground, the satellites are fed with accurate information about their position from ground stations. They then broadcast this orbital information, which is picked up by receivers aboard the ship. A small shipboard computer fixes the ship's position from the data.

Modern vessels carry extremely advanced apparatus, almost inevitably based on solid-state electronics, to enable them to work out their position with high accuracy, using signals from beacons, satellites, or their own equipment. In the latest type an aircraft's course is plotted by computer on a moving map so that the aircraft's position is always at the centre point of the display, enabling the pilot to keep continuous watch on his position. The information required is derived either from the ground radio beacons, or from self-contained navigation aids such as doppler radar, or inertial systems. The doppler unit sends out radio signals in front of the aircraft and behind, and works out the ground speed and drift from the differences in frequency of the reflected signals. The inertial system is based on a rapidly spinning gyroscope which holds its position in space, providing a fixed reference against which a computer works out changes in the craft's position and speed. It first saw service in the German V2 rocket bomb and then in ballistic missiles, and has come to be applied in ships, submarines, military aircraft and, from the mid 1960s, civil aircraft. Now it is in service with at least a dozen airlines, and is fitted as standard in jumbo jets. Virtually replacing the navigator, it can not only tell the pilot precisely where he is at any given moment, but also the time and distance to his destination. Connected to the autopilot, the inertial unit actually steers the aircraft. Such systems can compute the position of an aircraft to between 3 and 7 km (2 and 4 nautical miles) an hour, or of a ship—since the system is equally useful there—to better than 2 km (1 nautical mile) a day. This relieves the pilot of much effort: already airliners are flown by autopilot for perhaps 80 per cent of the flight. In a ship, the automatic helmsman steers better than a man for much of the time that the ship is at sea. The benefits have been felt in space, too; inertial guidance systems pointed the Apollo spacecraft to the moon.

While the radio beacons are certainly successful 'aerial landmarks', they cause crowding in the aerial highways. Because aircraft fly from one beacon to the next, they tend to be concentrated along quite narrow air lanes. It has led to claims from pilots that the vast ocean of air space has been converted into highly dangerous canals of limited capacity by beacon aids. The newer, self-contained systems of the moving-map type, however, allow planes to navigate for themselves, picking their own most direct routes and avoiding aerial traffic jams on existing air corridors. Radar sets warn aircraft of natural hazards—like bad weather—and of other aircraft. Marine vessels are warned of natural hazards and other ships. One of the latest types of marine set has an electronic memory that stores information during normal use and recalls it when required. This provides the navigator with a radar picture of the tracks of all vessels within range, in either true or relative motion, and allows the captain to see what would happen to the picture in the next few minutes if he were to alter course or speed by given amounts. Growing numbers of aircraft are installing the moving-map type of display. However, many ships are not as well equipped as they might be. From 1958 to

1967, on average one ship of over 500 gross tons was lost every three days worldwide. It is probable that 60 per cent of the losses would not have occurred if the ships had had better navigation equipment. In certain areas such as the waters between Britain and the Continent problems are acute. Half of the world's collisions at sea occur between the Elbe and the English Channel, chiefly because of the extremely high density of traffic passing through the Dover straits.

As we have seen, aircraft face similar hazards around airports. But aircraft are especially at risk at the moments of their transition from one medium—land—to another—air. The worst moments are take-off and landing, and the risks are higher when there is fog, snow, or rain about and when visibility is poor. Although research teams have been able to demonstrate automatic landings (which bring the plane down safely without the need for the pilot to see the runway) for more than thirty years, the problem has been to make this normal everyday practice. The world's major airlines are working towards the fully automatic landing, in which the aircraft is flown down by computer, guided along radio beams whatever the weather without the need for the pilot to touch the controls (see Diagram 8.3). Normally this requires that airports should be equipped with radio beacons of greater accuracy and shorter range than the other beacons we have referred to. Because they are expensive they come

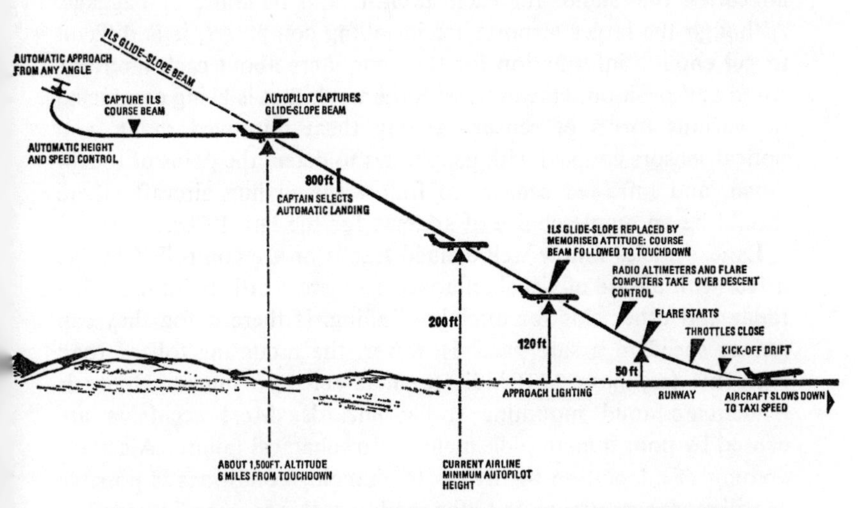

Diagram 8.3 Sequence of operations during an automatic landing (Elliott Flight Automation Ltd)

first to major airports in industrial countries, though they are just as vital at the minor airports at the other end of the flight. They represent merely one of the many location and communication systems that the airport needs. Improved radar plays a dominant role: for years, air traffic controllers have kept track of aircraft by watching radar 'blips' that told them the direction in which the planes were moving, but not how high they were or how fast they were going. In the newer type of radar, each blip is tagged by a computer with a cluster of continuously updated symbols that moves with the blip and conveys this information.

One possibility for bad-weather landing is to keep the plane under the pilot's hands and project the information about position in symbolic form on the windscreen—a 'head-up' display. Systems of the kind have been adopted in military aircraft. The opposite is to have a television-like screen in the cockpit, carrying the required data, with perhaps forward or downward radar views superimposed —a 'head-down' display.

There could soon come the day when landing, and the direction of aircraft once they have landed is fully under computer control. In the automated airport, aircraft would be guided in by computers that assume control some 24 km (15 miles) out from the runway. The control would continue after landing to govern such things as allocating the 'stand' for each aircraft, and handling of baggage. Although the larger airports are installing computers, it is difficult to get enough information for the computers about each aircraft's speed and position. However, with the work that is being conducted on various forms of sensor—among them improved radar, and optical sensors coupled with geophones to detect the point of touch-down, and infra-red sensors to follow the taxiing aircraft—there should be an ample choice of systems for the late 1970s.

Land vehicles follow well-defined tracks or are controlled by the driver with the aid of visual reference, so there is little point in having radar and other aids for direction-finding. If there is fog they can reduce speed to a safe pace. However, the mounting toll of road accidents (see page 278) indicates that here too human ability is inadequate amid mounting traffic hazards. Most accidents are caused by poor human judgement not mechanical failure. A certain amount of adaptation by drivers to changing conditions is possible as still more cars stream on to the roads, yet this must be limited. The general rule seems to be: the more vehicles the more accidents. With traffic volumes mounting—for instance, in the United States the numbers of cars, lorries and buses are expected to rise from 104 million to 158 million in the next twenty years—the prospect is grim.

So far, those concerned with car safety have concentrated on protecting people after an accident. Little has been done about preventing accidents by automatic guidance. One type of system, which would be a boon in fog, gives routeing directions at road junctions. At the start of the journey, the driver puts a five-letter code denoting his destination into a console in the car. As the car nears a junction the code is transmitted to a roadside unit that processes it and sends the route directions back in a few milliseconds. The instructions are displayed on a panel in front of the driver, or perhaps in a 'head-up' display. Of yet greater benefit to safety would be devices (incorporating some form of car-borne radar) that would keep each vehicle the correct distance behind another according to their speeds, and that would ensure that no vehicle would try to overtake another if there were danger of collision with an oncoming vehicle. Such radar systems are under development. But no doubt the motorist will fight to keep them off his own car.

Transport at Sea

At sea, speeds are limited by the resistance of the water to not much more than 27–37 km per hour (15–20 knots) for ordinary vessels. The sole way of increasing speeds drastically is to do as hovercraft and hydrofoils have done and become a hybrid partially (or wholly) out of the water, though still using the water for support. Naturally, they need a higher power–weight ratio to achieve the higher speeds. These craft can cruise at about three times the speed of an ordinary boat, and, given intensive development to produce them in larger sizes, could well offer fast, attractive cargo services over long distances. However, as they represent a completely new mode of travel they are still hedged about with economic uncertainty. No one has yet built a large enough craft to demonstrate what their costs would be in large sizes. Theoretically, 5000-ton hovercraft could cross the Atlantic in two days and nights at 150 km per hour (80 knots), but at present the largest British craft are 165-tonners plying across the English Channel with tourists and their cars.

The alternative is to make ships bigger to reap economies of scale, and more automated to lower the costs of the crew. Ships have been getting larger for about seven hundred years, at first slowly, then with a burst as iron, and then steel, replaced wood for construction, and steam and diesel replaced sail for propulsion. Oil companies are among the most anxious to reap these economies, as oil and its derived products are easily handled in bulk and have to be transported great distances from oilfields to market areas (see page 128). Tanker sizes have therefore increased rapidly in the past few years. In the late 1940s a vessel of 28,000 tons deadweight was regarded as a supertanker. (Deadweight, a measure of the ship's cargo capacity, is the difference in its displacement laden and light.) In 1959 and 1960 came tankers of over 100,000 tons. In 1966 the largest ship afloat was the *Idemitsu Maru* of 210,000 tons, yet 1968 saw the introduction of the first ships over 300,000 tons—six Gulf Oil tankers that carry oil from the Middle East to a storage and transhipping terminal at Bantry Bay, Ireland, for distribution by smaller tankers to refineries. Even with the added costs of transhipment and the capital investment of more than £10 million in the terminal, these ships will take crude oil round the Cape of Good Hope at about half the cost of shipping it through the Suez Canal in 50,000-ton tankers. Now ships of almost 500,000 tons are being constructed, and within ten years the million-ton tanker is feasible.

With advances in technology and design it has become cheaper to build and operate big ships which are able to incorporate more advanced equipment. One modern, highly automated supertanker, with a crew of perhaps thirty, can carry the load of six tankers of the 1940 vintage, whose combined crews would number some 245 men. Hull designs have been improved with the aid of computer-generated drawings. The bulbous bow, for instance, means an increase of about 3 per cent in average speed, and so reduces by an equal amount the length of time the ship spends at sea.

However, jumbo tankers, like jumbo aircraft, can bring awkward problems. Not many drydocks can accommodate these huge vessels, nor for that matter are there many ports which can. With their 24 m (79 ft) draught, fully laden, they have to choose their routes and their berthing points carefully. Higher standards of charting and navigation have become essential as tankers move with dangerously small margins of bottom clearance. Special facilities, often offshore, or in a deepwater port, have to be provided for loading and discharging the cargo of millions of barrels of crude oil quickly and cleanly.

Manoeuvring poses problems, too. When a large tanker is cruising ahead at 30 km per hour (16 knots) it may take 40 minutes to coast to a halt after the engines have been stopped, 9–16 km (6–10 miles) further on. Even in a crash stop, in which the engines go from full ahead to full astern as quickly as possible, it can take 12–13 minutes and about 3 km (2 miles) to bring it to rest. A giant tanker moving at more than 0·9 km per hour (half a knot) in dock can crush a quayside and hazard the ship.

The crew also have to adjust to the new conditions as automation aboard ship demands new skills of them. With few exceptions, existing ships are manned in a way that has been traditional almost since the early days of the steamer. A conventional ship, with a captain and a crew of say sixty or seventy, has served the merchant fleets of the world very well for fifty years. However, shipowners wishing to reduce crew costs, which typically account for one-third of ship running costs, are specifying automated systems for communication and control: automated engine room controls enable a tanker to be 'driven' by one man from the bridge with the engine room entirely unmanned. Men are only needed to keep everything in good repair, and so apart from maintenance their physical work is much reduced. The old divisions between the deck, engine room and catering are breaking down: the new seaman must be a better-trained professional, able to turn his hand to different tasks. It may be that in future a small 'pilot' crew will take a ship out to sea from port and then return to land, two or three remaining to conduct the ship on its voyage: at the other end another small crew will board and bring the ship into port again.

A further radical influence is coming in the handling of cargo, affecting life at sea and in ports. That is 'containerization', a jarring word with profound overtones. The basic concept is simple: carry all the odd-shaped packages that make up existing cargo in one large standard box called a container, and move that around from ship to shore and back. The containers go equally well on ships, lorries, trains and even aircraft. That means the operators can afford expensive mechanical handling equipment at the interface points between one type of transport and another to tote all the standard containers; it was impossible to arrange before because all the packages were different sizes. Productivity leaps; ships are loaded and unloaded quicker; and because they spend less time in port they can make more voyages and carry more cargo in a year. Fewer men are required to handle the cargo at the ports and elsewhere—maybe only one-tenth of the former dockers are needed—though they require higher skills to handle the new equipment. Hence a given volume of goods can be carried by fewer container ships, perhaps only half the number of conventional vessels. The sea carriers are faced with the prospect of scrapping existing fleets, re-equipping and merging with others if they cannot do so out of their own reserves. Irrespective of the way that they, and the men's unions, come to terms with these changes, the 'container revolution' is gathering force. In the first half of 1969, 40 per cent of all general cargo movements between Europe and the United States North Atlantic coast were containerized compared with 32 per cent in 1968. On the United States/Pacific route, the figure rose from 13 per cent in 1968 to nearly 40 per cent in the first half of 1969.

Diagram 8.4 shows the container chain in its essentials.

The concept only makes sense when the containers can be carried unopened all the way from departure to destination, and so common standards and practices are essential. Each part of the transport chain has to be able to handle containers from any state just as well as any other. There must thus be international agreement about such matters as the container dimensions, where the lifting points are on each and the type to be used, and accompanying documents and frontier clearance.

In fact, transport—whether of goods or people or ideas—cannot but have a unifying influence upon international affairs and practices, spreading common forms of technology across the world. It brings a standardization that irons out local, and national, differences: there

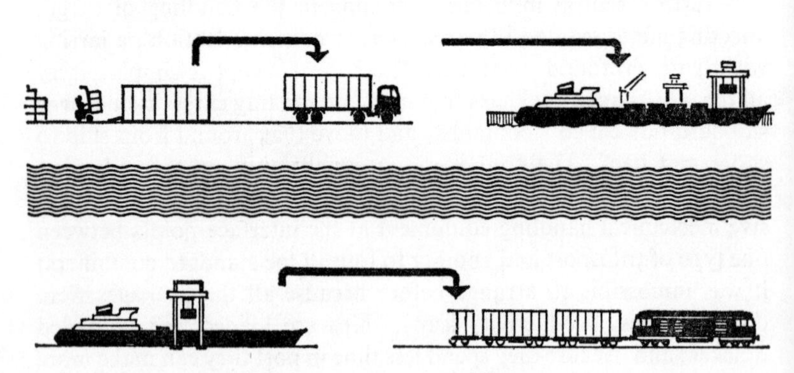

Diagram 8.4 The container transport chain

is nothing so standard as an international airport, a commuter train, or a mass-produced car. The need for common technical standards and for international controls and regulations on the movements of freight and passengers will exert a strong influence on political thinking in future years and on technology itself.

9 · The Hostile Void: Technology in Space

The Space Race

Space exploration represents the finest flowering of technology, or the most expensive and dangerous form of outdoor recreation ever devised, depending on one's point of view. There is no denying, however, that when the Soviet Union lofted Sputnik 1 into orbit in 1957, technology assumed a new dimension. It was the first time that Man had been able to hurl projectiles far enough from earth's gravity for them to go into orbit. It was not long before Man himself followed. On 12th April, 1961, Yury Gagarin, the little Russian cosmonaut, achieved the honour of being the first man in space, circling the earth in 89 minutes in his spaceship Vostok 1.

Technology had once again enabled Man to add to his long list of achievements, to grasp another opportunity of extending his power over the environment. It was the first time in nearly two million years that an earthbound race had been able to escape from the planet's gravity. It was an escape—albeit a short one—from the constraints of normal existence.

The event sparked off the greatest technological competition the world has ever seen outside war as the United States and the Soviet Union unremittingly pursued their next target, the moon. Their race has stimulated technology to an unprecedented extent, through massive flows of manpower, money, materials. By the start of 1970 the United States had spent an accumulated total of $44 thousand million on its space programme, and at the peak of activity in 1967–68 the National Aeronautics and Space Administration (NASA) employed 400,000 people, among them 5 per cent of the nation's qualified scientists and engineers. The American programme involved more than 20,000 industrial companies and 120 universities and laboratories. The total expense and effort in the Soviet Union can not have been very different. Overall the two nations must, so far, have spent something approaching $100 thousand million to get to the moon.

This enormous effort has enabled them to achieve in a few years results—in the form of new materials, guidance systems, medical knowledge and methods of managing complex projects, among others—that would otherwise have taken far longer to achieve, or might never have been achieved at all. It has done wonderful things for the technologies relating to jet propulsion, computers and miniature electronic components, for telecommunications, high-vacuum engineering, cryogenics and other technologies, mainly in the physical realm. NASA likes to point out that technology in general benefits from the 'spin off' of ideas, skills and devices created for the space programme. Hospital respirators, for instance, benefit from the principles used in the air-supply system developed to give astronauts on the moon a controlled atmosphere. An air-sampling device designed to collect extra-terrestrial particles at high altitudes can easily be adapted for air-pollution research. Resistant materials developed for heat shields on spacecraft are used as non-stick coatings on cooking pans. And so on, with many other materials, devices and techniques.

Then there are the advantages derived from the use of satellites. From their lofty perches they can, among other things, relay broadcasts over wide areas of the earth, spot hurricanes, distinguish mineral reserves and ocean currents, and discriminate between areas of diseased and healthy vegetation.

However, these are more excuses than reasons. The pity of it is that many of the goals could almost certainly have been reached for far less cost without going into space at all. It still costs anything between \$500 and \$1000 to put half a kilogramme of payload into a low orbit around the earth, and around \$14,000 to send a kilogramme to the moon. The most telling reasons for space exploration are curiosity and fear: curiosity about the origin of the earth and its neighbours in the solar system, and fear of the military advantage that a pioneer space nation could gain over others.

The space race has many other untoward consequences. Let us look at just two of them. One is the distortion of the educational system to train so many young people in the physical sciences, because those are the skills needed in the space and defence programmes. Research, too, has been directed to space and military inclined ends. Related technologies have blossomed in the United States and Soviet Union, and to a lesser extent elsewhere. Vast complexes, embracing government agencies and private companies have grown up around such programmes: in the United States, the military-industrial complex (which includes the space industry) is a recognized, if not well-defined, entity.

The other consequence is due to these two nations being the leaders of their respective world camps, as much in trade as in technology, so that other nations have tried to emulate them in space and compete with them in trade founded on the space-oriented technologies. The whole of the industrialized world's pattern of education and research has been distorted in the same directions; while many advanced countries aspire to put up some kind of a show in space. Thus after the Soviet Union and the United States put satellites into orbit, France felt it had to prove its prowess, and then Japan, too (in early 1970). China followed in April with a 172 kg (380 lb) satellite that orbited the earth in 114 minutes, playing a recording of 'The East is Red'. India has been investing heavily in space research. No doubt there will be others. Where they cannot stand the cost of going it alone, nations have to opt out of space activities altogether or else co-operate, as the European states are trying to do.

But there is mounting opposition to spending such colossal sums on, and devoting such massive effort to, satisfying curiosity or furthering military goals, however glorious they are for pure technological expertise. Many believe that it is mistaken to pursue projects that produce so little direct return for mankind when the world suffers a myriad other ills, as we point out in other parts of the book. The journeys to the moon, and the presentation of plans for the future American space programme, were undertaken amid an intensifying chorus of criticism. In the Soviet Union, though, such criticisms—if they were voiced at all—remained inaudible. But there is no doubt that they were aware of the attitude in the West. 1970—the target year set by President John F. Kennedy in 1961 when he pledged the United States to put a man on the moon within the decade—marked the turning point in the American space programme. The United States Government savagely cut back expenditure on the space effort, with traumatic effects on the whole aerospace industry, which had previously waxed fat on the generous funds.

The Future in Space

The first decade of space exploration was dominated by one goal—to get to the moon and back. That was achieved by American astronauts in July 1969. The future has no such compelling aim. The chief targets are those of consolidation: of constructing orbiting space stations, of making the moon habitable, of sending probes out to other planets, and even round the entire solar system. Man is aiming to make space travel safer, simpler and cheaper and to become more at home in this new, infinite environment.

Plans made by both the United States and the Soviet Union for the 1970s and beyond envisage space stations orbiting the earth, in which men could live and work for long periods. Then there are re-usable space shuttles to travel from the earth into space—say to the space stations—and back, landing on ordinary runways; space tugs, to carry men and materials about in space, to place or recover unmanned spacecraft in orbit and move cargo—a maid of all space work. Other objects include the landing of unmanned and, later, manned craft on Mars: and voyages by unmanned probes into the furthest reaches of the solar system. For the late 1970s, the United States plans to dispatch two such automated probes on a Grand Tour of the solar system. One would wing its way from earth close past Jupiter and Saturn to Pluto. The second, two years later, would fly past Jupiter and Uranus to Neptune. In each case the craft would be sped on by the gravitational field of each planet. This epic, nine-year journey would have to be done at that time because the planets will be aligned then to a closer degree than they have been, or will be, for a long time.

In addition, both nations will be consolidating their positions on the moon. The United States may build a moonport there, or several small lunar base camps from which exploring teams could rove about in specially designed vehicles.

Space stations provide a safe, well-equipped base from which to push on with the exploration of the unknown. The United States has studied a wide range of space station designs but has preferred to perfect first the equipment and the rendezvous and docking procedures needed for the moon venture. The Russians, however, have always professed to be less interested in landing a man on the moon (though they have landed machines) than in building orbital space stations from which they can make future space flights.

The first Russian space station, Salyut, was launched in April 1971, and boarded seven weeks later by three Soviet cosmonauts who rode up in spaceship Soyuz 11 and docked their craft with the Salyut. Together the two space ships formed a large 'celestial laboratory' about 20 metres (60 feet) long and weighing around 25 tons. The cosmonauts spent twenty-four days conducting an extensive space research programme before returning to earth. The flight ended tragically, though, when the three cosmonauts were found dead after an apparently flawless return journey to earth.

In spite of this tragedy, the United States is still intending to launch its first space station in mid 1973. This would be an adapted third stage of a Saturn 5, the massive rocket that carried many astronauts on the first leg of their moon journey (Diagram 9.1). Instead of fuel it would contain crew quarters, a laboratory and a working area. The plan is that a three-man crew in an Apollo spacecraft will rocket up from earth, moor their ship to the orbiting station, climb inside and spend a month or so in it. If the experiments prove satisfactorily that prolonged periods in space under these conditions are not harmful, the way will be open to constructing larger stations and sending crews up on a rota basis to man them continuously.

From such orbiting platforms astronauts could take observations of earth or other objects in space around them, and launch out on voyages to the moon, earth, or even Mars. A number of objects are better studied from space: the formation of weather in the earth's atmosphere, for instance, enabling more accurate forecasts to be prepared; the radiations and high-energy particles coming from space, and the filtering effect of the atmosphere on them; the earth's magnetic field and the upper layers of the atmosphere; problems of radio, telephone and television communication by satellite; methods of surveying from space; and military surveillance. A few permanent orbital stations, it is said, could provide as much information on these subjects as hundreds of satellites. The space station could also provide a base for emergency rescue—if, for example, astronauts were trapped in orbit in a space capsule whose retro-rocket had failed to fire.

Further, it could act as a factory for space vehicles. The component parts would be sent up separately and assembled at the space station. That would be a decided advantage with the planned inter-planetary vehicles for manned exploration. The moon rocket for the Apollo missions, Saturn 5, was bad enough. It weighs about 3000 tons, and hence needed engines developing that much thrust to raise it from the ground (see Diagram 9.1).

However, a launcher for Mars would weigh between 6000 and 28,000 tons on earth, and a Venus launcher would be heavier—say 9000 to 42,000 tons. Even if rockets with such huge thrusts could be made, the cost would be astronomical. Such programmes would almost certainly be beyond the capability of one nation, however

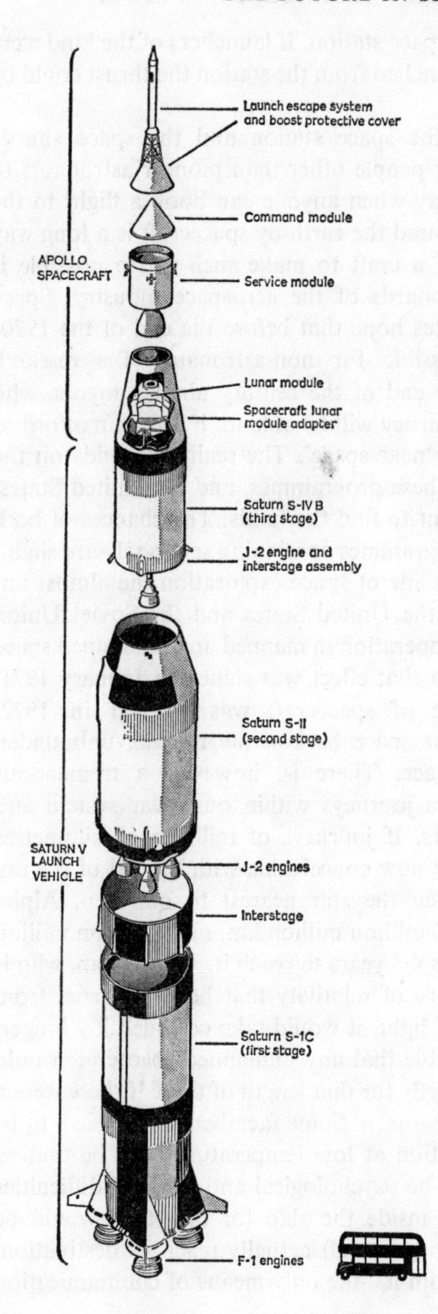

Diagram 9.1
Apollo Saturn 5 assembly, with a double-decker bus for comparison

rich, without the orbiting space station. If launchers of the kind were assembled in space and launched from the station the thrust could be much lower.

Still more important, the space station and the space shuttle would make it feasible for people other than pioneer astronauts to journey into space. The day when anyone can book a flight to the moon or take a journey round the earth by spacecraft is a long way off, but the forerunner of a craft to make such a trip possible is already on the drawing boards of the aerospace industry. Space officials in the United States hope that before the end of the 1970s space travel will be possible for non-astronauts like research scientists, and that by the end of the century almost anyone who wishes to go on a space journey will be able to, if they can afford it.

These are the plans for 'near space'. The reality depends on the money that is devoted to these programmes, and the United States, for one, is proving reluctant to find the funds. The chances of both American and Russian programmes continuing separately are slight. But early in the second decade of space exploration the almost unbelievable happened, and the United States and the Soviet Union started to discuss more co-operation in manned and unmanned space activities. An agreement to that effect was signed in January 1971. Accord on joint docking of spacecraft was reached in 1972. Beyond the realms of near space beckon the remote, unbounded vastnesses of limitless space. There is, however, a tremendous difference of scale between journeys within our solar system and those to other star clusters. If journeys of millions of kilometres within the solar system are now conceivable with the aid of the ion engine (see page 143), even the star nearest to our Sun, Alpha Centauri, is so far away (40 million million km, or 25 million million miles), that light itself takes 4·3 years to reach it. As for Man, who is assured by Einstein's theory of relativity that he is debarred from ever attaining the speed of light, it would take considerably longer.

It is virtually inconceivable that any unmanned spacecraft would be able to keep going perfectly for that length of time. If there were a crew, the problems would worsen. Some members might have to be kept in a state of hibernation at low temperatures and be thawed out near the destination. The psychological and physical difficulties with the crew cooped up inside the ship for that time would be formidable. Should such a spacecraft actually reach its destination, the radio or light waves from it—the only means of communication with earth—would take more than four years to reach earth again, as we have seen, even if they could be sent out with enough power to get there. While such journeys are rich fodder for science-fiction speculation, they are hardly credible, or creditable goals for technology. Nevertheless, Man will no doubt attempt to explore further and further into the galaxy, merely because it is there.

Behind Space Technology

Space exploration has proceeded by delicate steps. First, ballistic vehicles were fired to a predetermined altitude, with predetermined velocity, and positioning, and then coasted down to a preset target under gravity, using minor auxiliary power for control and guidance. The German V1s and V2s were of this type, and the success of both American and Russian space projects is largely based on the capture of the German scientists and engineers concerned in the closing stages of the Second World War. The development of intercontinental ballistic missiles (ICBMs) is another result (see page 308).

Next came the orbiting satellite (in 1957), which is boosted to about 90 per cent of orbital velocity, coasts out to orbit altitude, and then is kicked into orbit by a final rocket thrust. The moon spacecraft were ingenious, manoeuvrable, powered variants on this proven type. They employed rocket motors to kick them out of earth orbit, allowing them to coast to the moon, and then to enter, and leave, orbits around the moon, at will. On the return journey they did the same, only abandoning their rocket power—in a separate space capsule—just before re-entering the atmosphere. While the main rocket engine alters its velocity, an auxiliary system controls the attitude of the craft. This squirts out compressed gas through small nozzles to make delicate changes to the position of the spacecraft, allowing it to carry out the operations.

Finally will come spaceships, as yet on the drawing boards of industry, that will be designed to travel to other planets and back. These would probably be carried into space by another rocket, either entire, or in parts, to be assembled in space.

In order to put a satellite into earth orbit, it has to be given a velocity of about 29,000 km per hour (18,000 mph) in the upper atmosphere. Too slow, and the satellite is inexorably drawn back by gravity to a fiery demise in the atmosphere. Too fast, and it exceeds the escape velocity—about 40,000 km per hour (25,000 mph)—leaving the earth's gravitational field for an orbit round the sun.

Space engineers have to strike this critical balance hundreds of kilometres away from the craft concerned. To get the spacecraft into exactly the right orbit, with the right velocity and attitude at the correct time, demands extremely precise management, accurate working of all navigation and telemetry systems, and impeccable reliability from all components. On the ground, there are massive constructions—launch pads, control centres protected by steel and concrete shields, launch and service towers for the rockets, fuelling points, tracking stations to follow the rocket's progress, and all the other supporting facilities—to ensure that the rocket and its payload leave the earth in the peak of condition, and are monitored and controlled all the way (Plate 9.1).

The operation would have been impossible without computers, which are at the heart of all space launches. They carry out tests as the space vehicle grows on its assembly pad, sending out signals from a central point to the vehicle and receiving back the responses, which are then processed and stored on tape, displayed or printed out, as required.

In the last few hours before launch a 'launch computer' checks the performance of valves, relays, engine parts and so on, and relays the information to another control computer that searches in its memory for the corrective action and advises the computer on the pad what to do. After lift-off, a small computer in the Saturn rocket virtually controls the insertion of the spacecraft and the rocket third stage into the planned orbit. And naturally the manned spacecraft themselves are equipped with computers. So capable have computers and their software become that complete vehicle launches can be simulated and improved on computers without the cost and danger of carrying out the launch.

The spaceship is made in several sections, joined together for launch, and then progressively jettisoned. That allows the vehicle to derive the boost from the engines on each stage without being encumbered afterwards with the weight of the empty fuel tanks and the engines. The way in which the Apollo rocket is built up was shown in the previous diagram (9.1).

The penalty is that the more stages there are the heavier and more complex the assembly becomes, and the greater the thrust needed to lift the whole thing off the ground. Weight is always a critical factor, and research to discover strong, light materials here is more than usually important. The Apollo command module, for example, makes great use of metal honeycomb materials in its structure. During the Apollo programme engineers were having to pare grammes off the components of the lunar module in as many places as possible. It was because every kilogramme of load soft-landed on the moon requires about 120 kilogrammes of takeoff mass on earth, and also because of the delicate balance between the weight of the lunar module and the thrust available from its rockets during its descent and landing on the moon.

Much weight was saved in the command module, and therefore in all the other rocket stages that had to lift it into space, by the decision of both the space nations not to use retro-rockets in their manned craft to slow them down on re-entry. Instead, one end of the craft is coated with a heat shield of an epoxy-resin reinforced plastic, designed to burn away partially (ablation) as friction with the air heats it to around 2700°C. This dissipates the frictional heat as the craft decelerates in the atmosphere.

The rocket derives its thrust from the combustion and expansion of hot gases from ignition of fuel. Unlike the aircraft jet engine (see page 169) that gulps the air's oxygen so as to burn its fuel, the rocket engine has to carry both fuel and oxidant. Liquid hydrogen and liquid oxygen form one combination: these are prepared by supercooling (cryogenics, see page 74), and stored in very well insulated tanks on the launching site, and then in the rocket. Kerosene and liquid oxygen is another, and there are many more variants. These chemical propulsion systems are limited, however, in terms of the thrust they generate for the amount of propellant consumed in a given time, though they can produce very high thrusts per unit weight.

Better performance outside the atmosphere can be obtained from a nuclear rocket in which, say, liquid hydrogen is suddenly heated into a powerful propelling jet by a nuclear reactor, or an ion rocket, where the reactor's heat is converted to electrical energy which then accelerates the propulsive jet—in this case a stream of charged particles (ions). A given weight of ion fuel could last 100 times longer than the same weight of chemical fuel, and so these rockets ought to be able to propel spaceships on long interplanetary journeys with sizable payloads. It has been calculated that a typical ion-propelled spaceship could travel to Mars and back within 450 days, carrying 150 tons of payload.

Satellites

While no one doubts that satellites can obtain data about the rest of the solar system untroubled by the atmosphere that gets in the way of terrestrial observations, their value to earth is less clear. What can they do? The best example is the communication satellites. There are many now in orbit which relay signals for telephone, television, facsimile and other purposes from one ground station to another across the earth, without the cable system's need for a direct mechanical link between telephones. Theoretically, three could cover all the major inhabited areas on the earth if placed in a geostationary orbit 36,000 km (22,300 miles) above the equator, where they appear to hover over a particular spot on the earth.

In practice, it needs a great deal of expensive rocket power to put them in this wide orbit, and more than three are used, closer in, to improve coverage. The snag is that in this closer orbit the satellite appears to move—rise and set—so the ground stations that transmit and receive signals have to have steerable aerials, to track it across the sky, and that adds to their cost. Both the United States and the Soviet Union are developing this kind of satellite for military and civil purposes; and civil satellite consortia—which have to number one or other of these two nations among them, as they are the only ones with successful launcher rockets—are promoting more.

As satellite communications expand, so too will the number of ground stations. There are already nearly 100 of them, representing an investment of more than £100 million, and the number is expected to increase steadily in the foreseeable future.

Since December 1958, when the American Score sent back pre-recorded messages to earth, an imposing array of communications satellites has been put into orbit, starting with the large reflecting balloons of the early 1960s, which merely bounced weak signals back to earth. Now the satellites are 'active', that is, they pick up signals from earth, amplify them and then send them back. To do so they need a source of electric power—often solar cells—to drive amplifier and transmitter circuits. The first, Telstar 1, launched in July 1962, stopped transmitting after a year, having proved convincingly that the principle was sound.

Weights, and hence technical capability, have increased over the years. They will increase further beyond the current ones, of the order of several hundred kilogrammes, to craft weighing half a ton to a ton. The value of the added weight lies in the higher electrical power and electronic equipment it can carry, and the improved directional aerials which permit it to beam signals more accurately to specific areas on earth.

This kind of improved performance allows successive satellites to carry a higher number of channels and to serve a greater number of ground stations. For instance, Intelsat 1 (first known as Early Bird), the world's first communications satellite, was launched in April 1965 and relayed 240 telephone conversations between North America and Europe (Plate 9.2). Intelsat II, besides relaying telephone, television and data traffic, also acted as a communications support post in space for the Apollo lunar landing (see page 157). Intelsat IV, launched in early 1971, is designed to carry more than 5000 two-way telephone calls, twelve simultaneous colour television broadcasts, or any combination of different kinds of transmissions. It can point two 'spotlight' beams from two steerable dish antennae at any selected areas on earth, providing a stronger signal and greater channel capacity in areas of heaviest traffic. As satellite power is raised, the size of the receiving stations can be reduced. The process may reach a point in the late 1970s when the signals can be picked up by roof-top antennae similar to those already in use, allowing television to be beamed directly from satellite into homes.

The main advantage of the satellite is that it can see much more of the planet clearly than the earthbound human. Multi-band cameras, television cameras, infra-red and microwave sensors in the satellite can discriminate conditions on earth with far greater acuity than the human eye. The information can then be beamed back to earth tracking stations. A satellite can discern tidal currents, for example, or weather patterns that are invisible to human beings because they are too close to them. Several satellites are now regularly returning pictures of cloud cover for meteorological purposes. Ultimately, the information should lead to reliable, long-range forecasts by computer.

Further in the future, satellites will gather information about the earth's agriculture, forests, oceans, fresh water, geology and geography, assessing natural resources and making maps. The potential is there to detect diseased crops before their condition is noticed on the earth, pick out new bodies of ore, locate large schools of fish for ocean study and food supply, and detect fresh water in coastal regions where it may be seeping unknown into the sea. Exactly who will want to use data obtained at such great cost, which are thus most expensive, remains to be seen. When satellites of present size are joined by more ambitious space laboratories, it may become feasible actually to manufacture and process in space.

Problems of Space

Space technology has come a long way in an extremely short time. Although the Chinese were using rockets as a weapon in the early thirteenth century, in the late 1930s enthusiastic amateurs in several countries, notably Germany and the United States, were still experimenting with small home-made rockets and dreaming of sending men to the moon. They did not even possess vehicles which could travel far, or maintain their stability sufficiently to keep on an even keel or follow a preset course. Now, not much more than thirty years later, men have printed footprints and car bogey tracks in the moondust.

The successes to date have made us forget how astounding the achievements are, how extremely hazardous the whole venture is, and how slender the thread on which human life hangs in the void. Space is a hostile void, without air, with extremes of hot and cold, swept by showers of meteorites, by cosmic rays, and by bursts of solar radiation. The eruption of a solar flare could catch the astronauts without the thick shielding of the atmosphere, and would almost certainly injure them seriously. There is not sufficient shielding in the spacecraft to protect them: it would be too heavy.

This is no place for human life. Technology has to provide a micro-environment inside the space capsule—a simulation of the terrestrial climate—in which men can live and work. Everything the spacecraft requires must be carried with it, in particular the power supply. In manned flights especially it is vital that the sources, whether solar cells, fuel cells or batteries, work perfectly. Oxygen must be carried as pressurized gas, or supplied from the electrolysis of water. Exhaled carbon dioxide—1·1 kg (2·5 lb) per man per day—must be removed, chemically absorbed, or perhaps taken up by algae grown in the spacecraft.

A man needs just over a kilogramme (2·5 lb) of food, about 2 kg (about 4 lb) of water and nearly 1·4 kg (3 lb) of oxygen a day at normal temperatures. The food for the Apollo astronauts is either 'bite size' like a biscuit and complete in itself, or dehydrated in plastic containers and reconstituted by injecting water from a fuel cell. Waste products must be dealt with: the present practice is to store faeces, after adding bacteriological powder, in sealed plastic containers while dumping urine outside. Pressure inside the capsule must be maintained: if it were to fall sufficiently the astronauts' blood would boil. The temperature must be held at a tolerable level as the spacecraft moves from sunlight to darkness and the men exude the heat that is a consequence of their physical activity. Normally the spacecraft is rolled in order to turn a fresh side to the hot sunlight continually. There is also the largely unknown factor of the lack of gravity during space flight, and the heavy accelerations and decelerations imposed during take off and landing. Over periods of time in weightless conditions astronauts lose bone calcium, and unless a cure can be found or the process is shown to stop at some level, astronauts could suffer, particularly on longer journeys.

The spacecraft systems for oxygen supply to both the astronauts' spacesuits and to the cabin, the temperature and pressure controls, the power supply and all the other systems must thus work perfectly if disaster is to be averted. The fact that so many space flights have succeeded is a tribute to the technical efficiency with which the projects have been conducted.

The psychological problems are also daunting. The astronauts are confined in their cramped quarters, detached from the familiar environment on earth—suspended in the void out of human contact while the great globe that is their home hangs, like them, circling in the black void (see Plate 9.3).

The astronauts are the cream of the country's flying schools, and the low proportion of scientists among them, as against test pilots—albeit well-qualified ones—has been a source of friction. Between April 1959 and August 1967, NASA picked only 66 out of about 3000 applicants. In the stringent testing programme applicants were subjected to 56°C for two hours; held in reduced pressures; shut in dark cells for long periods to experience complete isolation; and spun in centrifuges; and they performed many trials with simulators and flight apparatus. They had to practise getting out of spacecraft in rough seas, and try out survival kits. All that in addition to the crews' practices for each flight. But they are human and they do make mistakes.

The performance of the equipment can also make or break a mission. On 25th January, 1957, the Americans' first Thor missile exploded after rising a short way from the launch pad. Similar fates overtook the first Jupiter rocket, which flew for 73 seconds, and the first Atlas missile, whose flight lasted 35 seconds. The first Vanguard rose 0·9 m (3 feet) before it fell back and blew up. In 1960 the first Project Mercury unmanned capsule rode skyward on an Atlas but the vehicle exploded after a minute. In the late 1960s, the first Intelsat III rocket destroyed itself after about a minute because of a control system fault.

Manned space flights have been bedevilled by minor technical faults. In some cases only good luck has carried a mission through, as when the Apollo 13 module suffered a near-disastrous explosion in one of its oxygen tanks.

Worst of all are the fatalities associated with the space programmes. In 1967 three American astronauts died in a fire in their command module during a launch rehearsal on the launching pad. One Russian cosmonaut died in 1967 when the parachute straps on his spacecraft, Soyuz I, fouled as it was preparing to land. And in 1971 three Russian cosmonauts perished during their spacecraft's re-entry to the earth's atmosphere, having completed a record 24-day orbital flight (see page 193). Faulty sealing of the spacecraft's hatch was thought to be responsible.

The main reason for these failures is the fantastic complexity of the machines. A rocket containing 300,000 parts—even if each one is 99·9997 per cent reliable—still has a mere one in 10 chance of complete success. Saturn 5 has 15 million parts, all of which must perform flawlessly. Throughout the vast Apollo programme the policy has been one of testing and proving all the smaller components, starting with valves, electronic components, etc. before incorporating them in larger ones, and building up those into larger ones, and so on, until the rocket is complete. This is most important, especially where it is a matter of virtually creating new technologies. Straightforward design is a direct extrapolation of previous designs. But space exploration has no proper precedents, and the testing programme for the spacecraft was the most rigorous ever given to any man-made object.

In preparing the components, great use is made of simulation, where conditions similar to those that the craft will meet are created on the ground in space simulation chambers (see Plate 9.4). These may subject the craft to humidity, acceleration, vibration, vacuum, heat and cold, so that failures can be found, diagnosed and rectified.

One spaceflight hazard—as yet unrealized—is that of colliding with a piece of space junk. By mid 1969 more than 3500 man-made objects had been placed in earth orbit, nearly half of which remained. There are currently more than 1800 objects circling the earth: about 1350 are junk, and some 450 are American and Russian satellites. It has been estimated that by 1980 there may be 5000 'dead' units and perhaps 1000 useful objects. By that time, there may be a heavenly rain falling to earth at the rate of ten of these objects every day, and there is a real danger of a high-flying craft such as a supersonic transport hitting a falling piece of debris with disastrous results. A much graver danger is that one nation's early-warning system will mistake a re-entering object of the kind as an attacking missile from another nation, and retaliate with missiles of its own (see page 308). The 'rain' has been falling for some years, and some of the 'drops' have not been completely burned up in the atmosphere, so that bits of old space hardware have dropped into South Africa, the United States, the Soviet Union and perhaps Britain and the Netherlands, among other places. A 10 kg (22 lb) section of a rocket nozzle, allegedly from a Russian rocket, crashed into a forest in Finland, for example, not long ago. Previously, a red-hot lump of the Russian Sputnik 4 had crashed into Manitowoc, Wisconsin, in the United States. A warning system to notify airline pilots of expected falling debris two months in advance was instituted in the United States in 1967, allowing commercial flights to be re-routed during the twenty or thirty minutes of danger. A warning and evaluation system at least as good is needed for military purposes.

The United Nations outer space treaty, though primarily designed with the purpose of banning weapons from space, also covers liability for damage caused by space debris on earth. It stipulates that those who launch, or help to launch, space vehicles are liable for any damage they cause. Moreover, all launching nations may be required to give assurances that their sub-orbital booster (rocket) stages will fall harmlessly into the ocean. It is a moral step forward, but leaves many practical problems untouched. If space launches continue at the current rate, there will be an urgent need for a space scrap-collecting satellite to remove unwanted objects from orbit.

While journeys into space may make nations—and mankind in general—proud of their technology, they do something more valuable in the present human situation. They allow Man to see his earth from outside, as never before: he can see how celestially unimportant he and his works are. The earth, for long believed to be the pivot of the universe, is just one planet circling round an average star (the sun), two-thirds of the way from the centre of a spiral galaxy. This galaxy, 100,000 light-years in width, contains 100 thousand million stars like the sun. Man's best optical telescopes disclose 10 thousand million galaxies comparable to his own in a universe of incomprehensible magnitude and impenetrable mystery.

Perhaps technology has a spiritual contribution to make after all, in showing Man such an awe-inspiring vision. Given the proliferation of galactic existence, it is extremely unlikely that homo sapiens is the only intelligent being in the universe. Some forms of life may be more advanced, others less so. Though technology cannot answer the eternal riddle of Man's existence, it serves to make him aware of his limitations and of the fact that he is probably not unique within this unbounded environment.

10 · The Last Frontier: The Oceans

The Oceans and their Reserves

Technology's last great frontier lies in the oceans, which cover more than 70 per cent of the earth's surface. Until now Man has treated them as either an inconvenience or a sewer. And yet they are as much a part of his environment as space. For thousands of years men have sailed across the waters of the world (possibly with rather more trepidation than they have explored land), have fished them, and have extracted salt from the brine. That has been as far as they wished to go. They have been content to treat this three-dimensional medium as essentially two-dimensional.

Not until in the past decade or so have underwater operations other than salvage or military activities been seriously considered. Man is on the brink of exploring and gathering the wealth of the sea and the seabed. While technology is now supplying the underwater vessels, diving gear, communications apparatus and underwater tools and propulsion, knotty problems of a different kind are arising. When the ocean depths were inaccessible, and thought to be of little value, no one laid claim to them. Now that their potential has been made clear, all kinds of claims—some wildly exaggerated—are being staked. In the last few years various nations have declared their sovereignty over the resources of the sea around their coasts: fishing claims have been extended as far as 320 km (200 miles) out to sea. Though not necessarily ratified by other nations, these claims are merely extensions of current practice in territorial waters. A far more vexed question concerns the ownership of the ocean deeps, the seabed and the rocks below. For instance, if valuable minerals are found in the underlying rocks beneath the ocean, how far can one nation extend its claim to them from the surface? Supposing that an economic way of extracting valuable metals from seawater were perfected, could the possessor claim a monopoly of all the reserves of that metal in all the oceans? These and many other legal problems are becoming more pressing as technology brings larger-scale exploitation nearer.

The best opportunities for exploitation occur in the comparatively shallow 'shelves' that surround the continents. These continental shelves are submerged under 20–550 m (65–1800 ft) of water, with an average depth of 133 m (440 ft), compared to the average ocean depth of 350 m (11,500 ft). Ranging in width from nothing to 1500 km, with an average of 78 km, they cover a vast area roughly equal to another continent the size of Africa. They contain the same kinds of resources as found on dry land such as oil and gas, sulphur, coal and minerals.

Much of the world's oil and gas supplies are extracted from the shelves through offshore boreholes, notably in the Persian Gulf, the Caspian Sea and the Gulf of Mexico. Eight per cent of the non-Communist world's oil came from this source in 1960; 17 per cent in 1970, and perhaps it will be over 30 per cent by 1980. In the North Sea, where abundant natural gas reserves have been recently found, drilling rigs and production platforms are operated at distances of up to 320 km (200 miles) out to sea.

Sulphur, like oil and gas, can be recovered from undersea reserves by boreholes. It is melted by superheated water piped down from the surface and forced up by compressed air in an elegant technical solution to a daunting problem. Coal, limestone, and ores of iron, copper, nickel, and tin are mined at present by tunnels driven out from under land in places as much as 25 km (15 miles) offshore. The economically feasible distance should increase to about 48 km (30 miles) by 1980 with new methods for rapid underground excavation; ultimately shafts may be sunk directly from the seabed if good deposits are found far from land. Another kind of deposit waits below. More than three-quarters of the continental shelves across the globe consist of loose sediments such as sand and gravel, tin ore, heavy mineral sands, oyster shell and diamonds. Some are already extensively dredged. Ten per cent of Britain's gravel for instance, comes from the sea. In the Far East, tin ores are dredged in waters around 45 m (150 ft) deep at great distances from the shore. In future they could be dredged by a machine working on the seabed.

More reserves lie farther out on the shelves themselves, on the upper parts of the slopes where the shelves start to give way to deeper formations, and on the submarine banks and ridges of the ocean floor. Phosphorite nodules—potential sources of raw material for phosphate fertilizer and other products—lie in deeper waters between about 30 m (100 ft) and 300 m (1000 ft). While large land resources last, these are unlikely to be utilized, except perhaps by countries without indigenous supplies. The best known phosphorite accumulations lie off North and South America, and South Africa. Manganese nodules are a more convincing prospect. They are lumps of manganese and other metallic compounds which have, by a natural process, settled on the ocean floor in concentrated deposits. They are commonly found in deep waters attached to rocks or other seabed objects, and are usually formed around a nucleus such as a shark's tooth or a stone. They are growing, like inorganic limpets, at a rate of about 1 mm for every 1000 years, by precipitation from the seawater. In spite of the difficulties of deep-water recovery, it seems inevitable that before long the world will need to turn to these sources as land-based reserves become scarcer. They are widely distributed, occurring in concentrated deposits in the Pacific and other oceans, and they appear to be plentiful enough to meet demands for manganese and other metals for many years to come.

More 'deposits' lie in undersea pools of warm, high-density brines in the middle of the Red Sea, that contain minerals in concentrations as high as 300,000 parts per million, ten times as much as in ocean water. They overlie sediments rich in such heavy metals as copper, zinc, lead, silver and gold. Similar deposits may be found elsewhere. It is probable that the technical problems of mining these sediments out of sight of land, and in deep water, will be overcome in due course. Problems of ownership are something very different.

Vast quantities of other potentially useful substances in the ocean deeps may eventually be exploited: calcareous ooze in the ocean floor sediments could be used for making cement; and siliceous ooze could be used in insulating brick, or as a filler (for instance, in plastics).

Just as immense are the reserves dissolved in the seas. Seawater probably contains all the natural elements, which have been leached out from the rocks since time immemorial. Yet only salt, magnesium and bromine are extracted in significant amounts. Seventy per cent of the world's bromine—used principally in anti-knock compounds in petrol—is extracted from the oceans. Sea salt has been extracted from seawater for thousands of years: the process can be traced back to Neolithic times. The food and chemical industries now use about 25 million tons a year. Nevertheless, it is only a beginning. The oceans with their dissolved solids represent the world's largest continuous ore reserve, containing almost unimaginable quantities of chemicals that may be measured in thousand million million tons, but with existing technology it is not yet economic to extract many of them. Reclamation from solid wastes may be preferred initially to reclamation from the oceans.

For many purposes, however, the fact that water contains all these dissolved substances is a nuisance, and much technological effort is devoted to removing them so as to obtain pure water supplies. Until the last few years, Man has been unable to use sea-water—representing 98 per cent of the world's water—because of the dissolved substances, and has had to make do with the 2 per cent of fresh water, most of which is, however, locked up in the frozen wastes of Greenland and the polar icecaps. Now new technological methods recover fresh water from the sea and other sources in large quantities (see page 216).

The recovery of pure water becomes more difficult and more costly as oceans and waterways become more polluted. Ever since human life began Man has discarded wastes into the oceans. At one time it was safe to assume that domestic and industrial waste would be safely and rapidly diluted, dispersed and broken down into harmless chemicals. Increasing population, together with rising industrial activity, has put an end to this comforting assumption. Coastal and inland waters are particularly prone to pollution, as they are closest to human activity.

Lake Erie, fourth largest of the American great lakes, offered a dreadful example at the beginning of the 1970s. Over the years it has been polluted increasingly by cascades of industrial wastes—from paper mills and rubber works, for example—and by sewage poured forth by cities around its shores, Detroit and Cleveland among them. Most of the rivers are themselves polluted. The Cuyahoga, flowing through Cleveland into the lake is so filthy that from time to time oil on its surface and methane gases bubbling up from the bottom catch fire. Bathing for 65 km (40 miles) along the lake shore is said to be hazardous to health. In summer, algae, which live on nitrogen and phosphorus compounds untouched by sewage treatment, collect in the western basin of the lake in a blanket over a vast area.

The open sea, too, is contaminated. Since tetra-ethyl lead was first added to petrol as an anti-knock forty-five years ago, lead concentrations in the Pacific have risen tenfold. Man-made radioactivity from nuclear bomb fallout can be detected in any sizable sample of water taken from anywhere in any ocean. Toxic DDT residues have been found in the Bay of Bengal, having drifted from as far away as Africa. Dangerous levels of mercury, possibly from industrial waste, have been found in tuna fish caught in the open sea. Pollution by massive oil slicks becomes ever more common as more offshore oil rigs are set up and the world's tanker fleet grows. Every few days oil is washed ashore on some part of the British coast. Chlorine-containing compounds, by-products of the plastics industry dumped in the North Sea, kill fish and the plankton on which they feed. The catalogue is a long one. Obviously, international action has to be taken and the much stiffer measures to control water pollution, introduced throughout most of the industrial world in the late 1960s and early 1970s, came too late to prevent drastic effects of a century or more of neglect.

Water—Supply and Demand

Man needs water: he is about 60 per cent water himself. Yet while in a primitive community a man can manage with 2·5 litres (a gallon) or so a day, in an industrial town the daily consumption can be as much as 220–270 litres (50–60 gallons) per person, of which about 160 litres (35 gallons) is for domestic purposes, and the rest for industry—in its widest sense.

The difference between the two totals lies in the effect of technology on life in the two communities. Thus, homes in a highly developed nation have their baths, water closets, showers and washing machines. In the fields, more water is used for feeding the animals, for watering crops, for washing down farm implements, pens and sties. In order to obtain 1 kg (2·2 lb) of meat from an animal it may require several tons of water, both directly for the animal and for the plants on which it feeds. If the natural supply can be supplemented artificially, then the land can bring forth more crops, and more meat.

But industry makes the main difference, with its copious thirst for water. The production of one litre (0·2 gallons) of petrol takes 10 litres (2 gallons) of water. A paper mill or large chemical plant may use of the order of 70 million litres (15 million gallons) a day, while about 325,000 kg (200 tons) of water are required to make one ton of steel.

Demand is rising rapidly. It is because of additions to the industrial family, such as the modern food-processing industries. Over 18,000 litres (4000 gallons) can be used in processing a ton of raw potatoes into one of the convenience foods such as crisps.

Irrigation is also on the increase; domestic use is rising because of rising standards of living, and, with an increasing population there are more consumers to supply. In 1939 the average daily consumption in Britain per person was 114 litres (25 gallons); in 1959 it was 205 litres (45 gallons); and now it is around 250 litres (55 gallons). But because of the factors we have mentioned, the overall demand in an industrial country such as Britain is expected to double before the year 2000, so that in thirty years water conservation projects will have to yield as many new supplies as all those constructed in the last hundred years (Diagram 10.1).

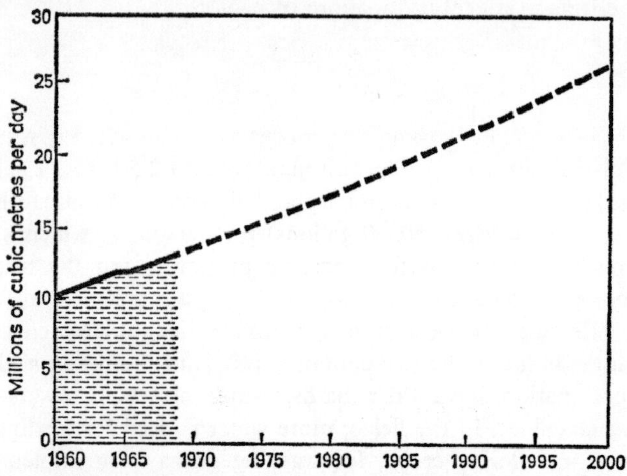

Diagram 10.1 Total water consumption in England and Wales, 1960–1969, with estimated requirements to year 2000 (Royal Commission on Environmental Pollution)

The chief opportunities for Man for obtaining more plentiful water supplies are by intervening in the natural hydrological cycle by which water falls from the clouds to the land, evaporates and runs into the sea, and is evaporated again. In many places rainfall is plentiful: around the world there are a number of areas with an average annual rainfall of more than 203 cm (80 inches), at one extreme, compared with the hot and cold deserts with less than 25 cm (10 inches). Where rainfall is sufficient, it may be technically feasible (the economics are another matter) to boost supplies simply by building reservoirs, constructing dams across river valleys and piping water from the impounded lake to the area where it is needed. In other areas, it is simply a matter of digging channels to divert river water. This method of irrigation is one of the earliest ways in which Man applied a form of technology to shape his environment. Around 3000 B.C. the first Pharaoh of Egypt had an embankment built to control the Nile's floodwater; and in China ancient irrigation systems, some 2000 years old, are still in use.

The principles remain the same, but now Man has the technical competence to build large dams in virtually any spot, to control the flow of capricious rivers, and to lead water for hundreds of kilometres across any kind of natural barrier, even through mountains, as in the Snowy Mountain scheme in Australia. Already many cities bring water hundreds of kilometres through pipelines from reservoirs in rural areas to meet their needs—and the needs will intensify as cities grow. Man has the power to organize water supplies on a grand scale—across whole continents if need be—just as he organizes supplies of electricity and gas and communications. A national water scheme has, for instance, been suggested for North America. Few schemes can match the Russian one to divert two Siberian rivers with the aid of dams and canals so that they can water arid lands in central Asia and then empty into the Caspian Sea. In theory, the river waters will make it possible to grow more crops, slow down the rate at which the Aral Sea is falling, and—incidentally—drain the west Siberian marshes, but the consequences of miscalculation on this scale could be disastrous.

Overall there is much more water in the atmosphere than falls to earth, while much of the water that does fall returns to the air again through evaporation from water surfaces and from the land, or through the leaves of plants. This suggests that, in order to increase the quantity of water available for human use, it would be worth trying to modify the weather, bring down more water from the skies, suppress evaporation, and limit unsuitable vegetation. All these methods are being tried (see page 269). Cloud seeding—dropping finely divided chemicals into clouds from an aircraft—can increase the rainfall below, though the results are at present slight. Evaporation can be controlled by means of storing water underground (the extent and nature of water-bearing strata is the subject of much research and survey work) and by spreading something, e.g. certain 'barrier' chemicals, on the water to stop it evaporating, especially on still waters in hot climates. Another method is to reduce moisture loss from the ground by planting certain types of vegetation, or in other ways.

However, as the scale and complexity of water development schemes increase, it becomes more difficult to predict what their effects might be. One new factor is that water supply, like other industries, has now to compete for land. Conflicts can arise and often do in heavily populated areas when land is deliberately drowned under man-made lakes, although the lakes can also be used for power generation (hydro-electricity), irrigation, wildlife habitats, and recreations like fishing, swimming and boating. Apart from the obvious drawback of flooding land, there can be other inadvertent effects. For instance, after the flooding of the Upper Zambesi River to form Lake Kariba, there was initially an explosive growth of waterweed that hampered navigation and the development of a fishery trade. Silting is another common and serious problem in man-made lakes.

But quality is as important as quantity, particularly as an increasing proportion of water has to be re-used, the partially diluted effluent from one community in higher reaches of a river becoming the raw supply for those downstream. Water is such an excellent solvent that in nature it is never pure in any case. Falling rain collects dust and dissolves oxygen, nitrogen, carbon dioxide, ammonia and ammonium nitrate. Near the sea it contains salt, and around urban areas sulphur dioxide and sulphuric acid. In the ground, water attacks rocks, decaying organic matter, and any soluble substances in the soil—for example, calcium and magnesium compounds that make the water 'hard'. Water from a natural source invariably contains bacteria, too. To this 'natural pollution' Man is adding a growing amount of his own. The waste waters from home, laundry, steelworks, brewery and chemical plant, contain dissolved or suspended impurities that must be removed before the water can be re-used.

Underground water, or water from fresh sources stored in covered reservoirs, can often be used simply after sterilizing it with chlorine or ozone. Waters taken direct from rivers, and appreciably coloured waters from upland reservoirs, need more extensive treatment and generally have to be filtered first through sand beds to remove harmful organisms. Where the water is coloured and contains suspended impurities, a coagulant chemical like aluminium sulphate is usually added to precipitate impurities as a 'floc' which is removed in settling tanks and sand beds. Lime or other chemicals may be added to control its hardness.

Though the importance of such treatments has been appreciated since the nineteenth century, it was only the technological evolution of steam-driven water pumps, and mass-produced water pipes of cast iron, in the same century that made it practicable to supply large industrial communities with piped water. The combination of biology, chemistry and civil engineering applied to water supply and treatment makes it possible to produce massive quantities of extremely pure water at low cost, at least in the developed countries. Elsewhere the situation is less impressive. Only 10 per cent of the world's population is believed to have piped water supplies, even now, although the provision of better supplies in developing countries, specially ex-colonial ones, has done much to bring down death rates among the population. In these days, too, the hazardous, crude natural world has a habit of breaking through the comforting technological veneer from time to time. In the early 1960s typhoid broke out in a fashionable Swiss holiday resort, carried by an infected water supply, and in 1970 the United States Government found that there was serious cause for concern about America's drinking water. Of the 1000 community water systems it studied across the nation, more than 400 were providing water of inferior quality, and it seemed likely that 6 per cent of the population were drinking water of a potentially dangerous quality.

Present large-scale purification methods have certain limitations: they are, for instance, unable to remove dissolved salts. Industrial plants often require very pure supplies—e.g. in high-pressure steam boilers, or sensitive processes such as the manufacture of transistors—which require special treatment. In the future, as water has to be re-used more frequently, both for industrial and domestic purposes, further stages may have to be added to water treatment plants at extra cost. When water is used merely for cooling in industrial processes, quantity is usually more important than quality. Thus a large power station takes in cooling water at rates of 270–370 million litres (60–80 million gallons) per hour, returning it after one circuit of the plant to the waterway. It is often possible within a limited area to take treated sewage effluent directly for cooling supplies. Irrigation can use this source as well. Even so, water has been cheap in the past, reflecting the ease with which it was obtained from waterways, boreholes, etc., and its price will have to rise as demand becomes more insistent and the cost of treatment—amid growing pollution—increases.

In drier places alternative sources of water have to be found to supplement the natural supply. One of this century's prime innovations is the discovery how to extract potable water cheaply and on a large scale from brackish or salty water. It has been common practice for ships to obtain fresh water by distilling seawater since the end of the nineteenth century, but only quite recently has technology made it economic to do the same on land on a large scale. The search for minerals and oil in arid, sparsely populated areas such as the Middle East has prompted much development on methods of obtaining water in such unpromising surroundings. Many small, mobile desalination units of the electrodialysis type (see page 217) are carried by mining teams, mineral exploration teams and similar groups.

One method of obtaining pure water is an elaboration of simple distillation. In commercial plants the process is mostly carried out under reduced pressure to lower the boiling point, and in as many as thirty to forty stages, in what is called multi-flash distillation (see Diagram 10.2). Some of the heated water 'flashes' into steam as it is injected into a chamber in which the pressure has been reduced, and pure vapour is collected on condenser tubes. Each chamber is at a slightly lower temperature and pressure, and the water is passed through them all. Sea water is re-circulated through the condenser tubes so that it picks up heat from the hot condensing vapour, and is then fed into the first chamber of the series once more. This means that it needs little heating before its temperature is high enough for it to be injected at the start of the cycle again.

Another method makes use of the fact that when brine freezes the ice crystals that separate out are almost pure. At the same time, pure

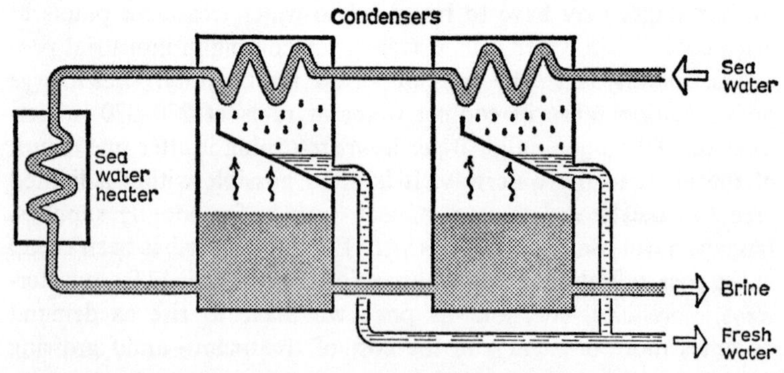

Diagram 10.2 Principle of the multi-stage flash distillation process

vapour is drawn off and pure water regenerated from the two together. Another, electrodialysis, is a suitable process for brackish waters. It clears impurities by passing a current through the water, which is contained in a series of tanks separated by membranes that allow different ions (the electrically charged particles of the dissolved solids) to pass through. The ions are collected at the electrodes, leaving the water between the membranes relatively free. A major advantage is that the plant is fairly cheap, especially in small sizes. Yet another method, reverse osmosis, relies on squeezing impure water through another type of membrane (it could be a plastics tube of a certain type) which allows the water through but holds back impurities. In theory it is simple but in practice there are tough problems associated with devising membranes that will stand up to the high pressure for long enough. A large part of the work being conducted on these processes throughout the world is devoted to designing cheap, efficient, reliable and long-lasting membranes. Other processes are possible, of course, such as the Israeli direct-contact, counter-flow method which eliminates the need for heat-exchange elements.

All desalination plants need a source of energy. Where sunshine is plentiful solar power is enough for small installations. Survival kits for marooned airmen who ditch in the sea generally contain a solar still to produce pure water in this way. Elsewhere, and in larger installations, a more substantial source is required. One neat way is to utilize the waste heat from a power station. Plants of the type have been installed in many parts of the world, notably the Middle East. A large plant, running on steam from an existing power station, is capable of producing 4·5 million litres (1 million gallons) a day of fresh water for drinking and domestic consumption, containing not more than thirty parts per million of solids, and some very pure water for the power station boilers, with not more than one part per million. It should be possible to bring costs down by linking the desalination plant with a nuclear power station. Several such schemes have been announced, but the one which appears furthest ahead at the time of writing is that at Shevchenko in the Soviet Union, near the Caspian. Its fast breeder reactor will generate 150 MW of electricity and the associated desalination plant will produce 150,000 tons of fresh water a day.

At its most adventurous, this concept leads on to the nuclear-agro-industrial complex, or nuplex. Its heart would be a nuclear power reactor, supplying waste heat to the desalination plant. Electricity would electrolyse water into hydrogen and oxygen, and the hydrogen would be converted into ammonia fertilizers in an associated chemical plant. Phosphorus fertilizers would be produced from phosphate rock, probably imported, in an electric furnace that would be powered by the reactor. Desalted water, and fertilizers, would be applied to the land for irrigation at just the right stage in the plants' growth and in just the right quantities. Crops would be specially bred for the man-made conditions, and would primarily be the very high-yield grains, such as the successful wheat developed for Mexico by the Rockefeller Foundation (see page 237), which would yield several crops a year with a small amount of water. No such complex has yet been built, though fairly detailed plans have been prepared. As it is, there are more than 680 desalination plants with a capacity of more than 113,000 litres (25,000 gallons) of fresh water a day in operation, or under construction, around the world. The growth in the market is projected at 25 per cent a year over the next ten years, so strong is the demand for water.

Working under the Waves

Before any exploitation can take place, more must be discovered about the mysterious oceans and their strange forms of life. Investigation from the surface is limited, although the technology brought to bear on the problem has improved tremendously in the past decade. Many data have been gathered, it is true, by lowering bottles, thermometers and other sampling instruments first on hemp ropes and later on wires or wire ropes. These measurements, and dredging of animate or inanimate specimens for examination, had, up to a few years ago, given Man his only clear ideas of the sea and the seabed.

Instruments have been refined considerably in recent years, especially through the application of integrated circuits; and exploration by dredging has been supplemented by underwater photography, closed-circuit television, sonar, electrical, seismic and other techniques. Sonar, the sound counterpart of radar, is a powerful tool for location and exploration. Sound is the sole kind of radiation that penetrates water to any distance and hence searching and telemetry done underwater usually employ sound. Many types of equipment are made. Doppler sonars (compare with doppler radars, see page 181) enable a ship or submarine to calculate its speed over the sea bottom. A sonar beam from the ship can trace a picture of the sea floor, with rock outcrops, wrecks, shoals of fish and similar features clearly displayed.

Knowledge of ocean currents is only now becoming fairly realistic, and far less is known about the deep and bottom currents, which cannot be measured except by specially equipped ships. The task has been simplified by neutrally buoyant floats (invented in the mid 1950s) which are fitted with a sound source so that they can be tracked from the ship as they glide with the current.

In fact, the depths prove to be far less settled than was formerly supposed. Currents of about 0·6 km per hour (one-third of a knot) are not unusual at any depth on the western side of an ocean. Some currents are much stronger. Particularly fierce ones are associated with underwater 'landslides' set off when a pile of sediment accumulated on the top of the continental slope—eroded material carried from the land—suddenly topples and rushes down into the deeps. Such material, travelling at speeds of as much as 97 km per hour (60 mph), can break submarine cables running along the slopes.

With the aid of all this apparatus, Man has probably learned more about the oceans in the past fifteen years than in the preceding 5000. However, while instruments are useful, there is so far no substitute for putting a man down there to observe and act. Not until the middle 1960s did the exploration of 'inner space' really get under way, with Man using technology to enable him to plunge into the depths. The opening of the 1970s saw shoals of underwater vessels, being lowered, driven, or tethered to the bottom; craft ranging in size from one-man 'sledges' like torpedoes, through work-vessels equipped with mechanical claws and other tools for underwater engineering, to 'houses' holding teams of ten or more men for periods of weeks (see Plates 10.1, 10.2, 10.3). These vessels, known as submersibles, conducted exploring aquanauts to depths as great as 2400 m (8000 ft).

The impetus for exploration has come mainly from the oil industry on the one hand, and military interests on the other. The military value of having underwater recovery vessels was proved by the United States when its manned submersibles recovered debris from the stricken nuclear submarine *Thresher*, and found an H-bomb that had dropped into about 800 m (3000 ft) of water off Palomares in Spain after an American bomber crashed there in 1966. With fleets of rival hunter-killer submarines stealthily stalking each other beneath the oceans, vessels for underwater rescue appear still more vital. The United States has two deep-submergence rescue vehicles which can be carried by giant transport aircraft to the port nearest to a sunken submarine, and then sail piggyback to the rescue site. Once there they leave their mother vessel, locate and 'mate' with the crippled submarine to form a watertight passage through which the survivors can be hauled to safety. These can operate in depths up to 1520 m (5000 ft).

A growing proportion of oil and gas has to be extracted from under the sea, though it is two or three times as costly to develop an offshore field as one on land. For the oil industry, manned submersibles can carry out a variety of valuable tasks such as inspecting and maintaining oil well-heads and other installations, conducting underwater surveys, finding faults in cables and pipelines, locating and identifying wrecks and salvaging. As the offshore search widens (see page 144), rigs are having to be constructed in wilder waters further out. The first platform for submarine drilling—as opposed to drilling in lake or marsh, which was done earlier—was built in 15 m (50 ft) of water in the Gulf of Mexico in the late 1940s. So rapid has progress been since then that there are now some 2000 platforms there, linked to the shore and each other by an estimated 8000 km (5000 miles) of pipelines.

At present, nearly all offshore oil production is from fixed platforms of which three-quarters are in water no more than 25 m (80 ft) deep, but the oil industry is resigned to having to evolve techniques for working in rougher, deeper water in the next few years, much deeper than the 100 m (350 ft) or so that represents current limits.

Though the technology is ready to build fixed platforms in 180 m (600 ft) of water, it is an expensive business and the trend is to drill from mobile rigs that can be moved from one well to another, spreading the cost of the drill structures over a greater number of holes, and then to construct the well-head gear (including a means of separating oil and gas) on the seabed—with remotely controlled valves and pipelines connecting it to central collecting points. The jack-up type of rig, whose massive legs stand on the seabed, and the ship-type unit have recently been supplemented by the semi-submersible type, which floats, part above and part below the surface, on large cylindrical columns to minimize the buffeting of waves. This has allowed commercial wells to be drilled in depths of 396 m (1300 ft) of water.

In order to work on such installations, workmen are often lowered in a steel capsule to a 'cellar' maintained at atmospheric pressure fitted permanently over the well-head. Electric power and air are piped to the capsule through 'umbilical' lines from the surface. The emphasis of technology is in fact shifting from having machines on the surface to actually working on the seabed. The world's first underwater storage tank for oil, capable of holding 500,000 barrels, was installed off Dubai in 1969, allowing tankers to anchor and load near it in deep water, saving the cost of pipelines to the shore. It has been suggested that oil drilling rigs, refineries and all ancillaries could be constructed on the seabed over the oilfield; and, going further, that nuclear power stations, explosives and other potentially hazardous plants could be built and operated underwater.

Recently the first seabed dredge has been put to work, pumping back spoil to the shore, and moving along the bottom on crawler tracks in depths up to 60 m (200 ft), unaffected by surface waves. In Japan, an underwater bulldozer, controlled and supplied with hydraulic power from a catamaran, has set to work digging, shovelling and making trenches underwater. Many other seabed machines are envisaged, to explore and to work on the sea floor in a similar way to those that move and construct on land.

Meanwhile the exploration continues. Underwater mariners have penetrated the ultimate ocean deeps in bathyscapes which being lowered from the surface, are less mobile than submersibles, to reach depths of as much as 10,750 m (36,000 ft) found in the Marianas Trench. It is the kind of pioneering venture which brings Man into contact with enigmatic forms of aquatic life in the abyssal gloom, a realm of near-freezing temperatures and enormous pressures. In order to make hulls which can withstand these conditions and are still light enough to return to the surface, special design techniques have had to be employed. So too have the strongest, lightest materials such as glass-fibre-reinforced resin, glass, aluminium alloys and new forms of steel.

Life in the Oceans

In the next few decades, Man will in all probability return in numbers to the deeps whence life first crawled millions of years ago. Divers may farm the continental shelves, raising fish in pens in the same sort of way that livestock are reared on land; they may also work on oil well-heads and repair machinery at depths below 300 m (1000 ft), while mining engineers prospect for reserves on the seabed in deep-diving submersibles. Machines themselves will crawl across the sea floor, scooping up samples, laying cables and pipelines, constructing underwater dwellings and even industrial plants. Scientists will study the ocean and its life in habitats tethered to the ocean floor. Families may take holidays in warm seas, living in pressurized underwater 'hotels' and going out for subsurface tours and fishing during the day.

With all these potentialities now being considered, it is strange that so much time and effort should have been devoted to exploring outer space and so little to the oceans—'inner space'—which have yielded a steady return for so long, in the form of the fish catch. Among the many possible reasons for this neglect is the fact that all human activities are normally carried out in air so that it was more natural for Man first to explore the atmosphere than the sea, in which unaided breathing and fuel combustion are impossible. Underwater exploration thus started with physical handicaps. In addition, it has lacked the kind of overriding goal that dominated at least the first decade of space exploration. There is also a kind of psychological handicap, surely, in the contrast between the light and airy heavens and the dark, cruel, impenetrable sea, where many mariners have perished.

While such underwater feats are distinct technical possibilities, the cost of implementing them will be heavy. There are many problems to be solved, concerned with keeping men alive and well in the hazardous environment—which is no less hostile than outer space—and doing it for reasonable cost.

The aquanaut like the astronaut is isolated in a strange world where his only link with the surface is through an umbilical cord carrying his communication lines and air supply: in some cases he may discard that, relying on self-contained apparatus. He may be in almost exactly the same 'weightless' state floating in water as an astronaut is in space; and indeed, American astronauts rehearsed weightless movements wearing space suits and floating in tanks of crystal-clear water (Plate 10.4). He is subjected to great pressures, numbing cold, a rapid decrease in sunlight down to absolute darkness, turbidity that may cut off visibility, even when there are artificial lights, and dangerous forms of marine life. Most important, he has to be supplied with oxygen, usually diluted with an inert gas such as nitrogen or helium, and his exhaled carbon dioxide has to be removed. Until the nineteenth century, if men ventured below the surface, it was for as long as they could hold their breath, apart from a few hardy souls who were willing to commit themselves to crude forms of apparatus—animal bladders, and rudimentary diving helmets or diving chambers—which were all that could be provided before then.

The problems multiply because his breathing mixture has to be pumped into him at a pressure equal to that of the water around him, and this forces gas into his blood, where it can cause narcosis. Nitrogen narcosis caused by normal air mixtures below depths of about 45–75 m (150–200 ft) can turn a man into a senseless 'drunk'. Replacing the nitrogen with another inert gas, helium, gets over the narcosis problem for deep diving down to about 450 m (1500 ft), where helium narcosis sets in, but it brings others. The diver talks in a Donald Duck squeak that makes communication with the surface difficult, although the helmet can be fitted with an electronic unscrambler to correct it. Moreover, helium dissipates heat much more rapidly than nitrogen so that the diver gets cold more quickly, and the suit has to be heated to keep him warm.

In spite of the difficulties, improving technology has enabled men to penetrate further into the depths. In the eighteenth century divers could work steadily at 33 m (110 ft). The invention by Augustus Siebe of a metal helmet fitted with adjustable air inlet and outlet valves in 1819 (a more advanced design came in 1830) allowed divers to go deeper, and by the early 1940s the limit had been extended to over 90 m (300 ft). After the Second World War it was again extended, to 180 m (600 ft) at sea, and around 300 m (1000 ft) in a simulator diving chamber in the laboratory. During the current decade, though dives are made deeper as time goes by, it is unlikely that divers will be used below about 610 m (2000 ft).

In the middle 1940s the increasing depth of offshore petroleum exploration made it economic to convert the experimental diving techniques into routine working methods. But until the invention of saturation diving it seemed that men's ability to live in the sea was extremely limited. The reason is that the diver has to spend a long time in decompressing as he returns to the surface to rid his tissues and blood of excess gases forced into them by the high pressure. If he returns to atmospheric pressure too quickly, these gases form bubbles that produce painful and sometimes fatal 'bends'. In order to avoid these, he can either come up gradually, waiting for several hours at specified depths as he comes up, or wait in a decompression chamber on board ship at gradually reduced pressure. In either case, it is a slow business. The greater the depth of the dive the longer he has to wait, and so the smaller the proportion of the total time that he can usefully spend working below. Roughly, a dive to 90 m (300 ft) requires thirty hours' decompression.

However, in experiments in the late 1950s the U.S. Navy found that if a diver stays under pressure for about twenty-four hours, at any depth, his body becomes saturated with the gases. He only needs to decompress when he comes to the surface, irrespective of how long he has been down. This discovery ('saturation diving') showed the way that underwater living might start. If divers could stay in a diving chamber in an atmosphere at the same pressure as the water outside, they could venture into the ocean as often as they wished, and would only have to decompress on their return to the surface. That was the starting point for many 'underwater living' experiments, with the United States to the fore. One type of diving apparatus consists of a decompression chamber on the ship's deck, and a diving chamber that is lowered into the water. Divers are lowered in the pressurized chamber to their work on the seabed, and then hauled up to transfer to the pressurized deck chamber after they have done their stint. A second team takes their place below while they relax in warmth above. Both teams—or several, if the job demands—are kept pressurized until the work is done, when they are all decompressed together in the deck unit. This allows divers to work for weeks at a time in depths down to 260 m (850 ft), as deep as most parts of the continental shelf. Alternatively, divers may be carried in a 'lock-out' submersible which has two compartments: one, for the pilot, is kept at atmospheric pressure, and the other can be pressurized. The divers travel to the work site, pressurize their compartment and go out through a hatch into the water. When they have finished their job they come back into this compartment, and the submersible takes them back for decompression in the deck chamber.

The other approach is to live actually in the sea in an underwater habitat; and since the 1960s larger teams have been spending days and even weeks in experimental habitats at increasing depths. These pioneer 'aquadwellers' may carry out salvage on wrecks, build structures on the seabed, test underwater tools and make excursions to greater depths. Above all, they register their physical condition and their reactions to prolonged spells in these strange surroundings.

Exciting though these experiments are, they are but experiments. The cost of providing reliable facilities underwater is at present prohibitive, and underwater habitats have to be supplied with power, air, communications and other services from the surface unless they are very close to shore. Thoughts of large numbers of people living underwater for extended periods are therefore little more than pipe dreams for the immediate future.

Man's chief interest in ocean life is of course its food value. There is already abundant life in the oceans, ranging from mighty whales to microscopic plankton, which Man has found a constant source of food since antiquity. Man's annual catch of these forms of life has been increasing fast. In the century from 1850 to 1950 it rose tenfold, at an average rate of about 25 per cent per decade. In the next decade it nearly doubled. At the moment it probably amounts to nearly 70 million tons, 90 per cent of which is finfish, the rest being whales, molluscs, crustaceans, etc. Present forecasts are that it will gradually rise to around 140 million tons annually, for the kinds of fish now being caught. It is disturbing that in the past few years the rising trend has been sharply interrupted, indicating that the cost of fishing, amid many over-fished stocks, may be rising faster than the returns it can bring.

At the present time, nearly all of the catch helps to feed Man, providing about 10 per cent of his total protein. Some is eaten directly as fish. Some is made into oil and meal used for food products, such as margarine, or for feeding poultry or pigs. It may also be turned into fish protein concentrate (see page 243). Just before the Second World War less than 10 per cent of the world catch was converted to meal, but by the end of the 1960s the fraction had risen to half.

In order to increase food production from the oceans, new fish resources have to be discovered, and known but unexploited stocks will have to be exploited as only a dozen out of 20,000 known fish species provide 75 per cent of the catch. Fishing methods and equipment will have to be improved and more fish cultivation undertaken. To avoid fish spoiling once they have been caught, handling methods and the processing and distribution of fish and fish products will have to be streamlined.

Even so, technical advances have already transformed fisheries from predominantly small-scale, local ventures to a modern industry. Large factory vessels, with facilities to process and freeze fish on board, served by catcher trawlers constitute the most mobile production units in the world, travelling thousands of kilometres to and from fishing grounds. A recently completed Russian factory ship, for example, carries a flotilla of catcher vessels, lowering them into the water at a fishing ground and hoisting them aboard again once their work is done. The individual catcher vessels can be equipped with a variety of modern aids, enabling them to find and catch the fish speedily.

These modern aids are needed urgently. The traditional harvest of the sea is increasingly elusive, because in many areas over-fishing is depleting stocks faster than they can replenish themselves. The threatened stocks include herring, cod, and ocean perch in the North Atlantic, some tunas in most ocean areas, and anchovy in the south-east Pacific. The traditional practice of a fishing fleet moving from an over-fished to a more plenteous area cannot continue indefinitely because there is but a limited number of new fishing areas. The North Atlantic and North Pacific are heavily over-fished areas, unlikely to yield more than a few free-swimming species in future. The Indian Ocean is a better prospect, however, since it is the source of only 5 per cent of the world catch, compared with the Pacific's 53 per cent and the Atlantic's 40 per cent, and it is not a biologically poor area. Nevertheless, most fishing grounds lie in the shallow waters over the continental shelves, and the vast areas of the deep ocean—though not fished to any great extent—are unlikely to be similarly productive. In some oceans, intensive exploitation has been limited on the one hand to species that occur in large concentrations—such as the anchoveta, fished in such large quantities off South America that Peru lands the world's heaviest fish catch, and the pilchard, found off the west coast of South Africa—or on the other hand to species with a high unit value, such as tuna in the open ocean and shrimp along many coasts. In future, many more available species will have to be caught and eaten, as the stocks of the currently popular varieties dwindle. In the meantime, moves have been made to curb over-fishing, and international agreements limiting the catches of whales and salmon were notable examples during 1970.

Man has always hunted fish: only now is he beginning to think of farming them widely, although the practice is of great antiquity in the Far East. Though difficult, if not impossible, for most free-swimming species, fish-farming is carried out successfully in many countries and in several instances on a relatively large commercial scale. Such immobile creatures as oysters and shrimps are among the easiest to handle. Fish may be raised artificially so as to overcome the high mortality rates in early months, and then returned to the sea, or they may be reared in pens (the 'broiler-house' approach, see page 237) in land-based enclosures. Some reared experimentally in the warm water discharged from power stations grow to mature size much more speedily: plaice have in two and a half years reached a size that would normally take five, and sole in eighteen months one that would take three to four years. There are problems in scaling up experiments, such as obtaining cheap enough food.

Other more spectacular ways of increasing the catch are the harvesting of certain plankton like the Antarctic krill—small shrimp-like crustaceans that are the main food of the baleen whale—and fertilizing the warmer surface layers of the ocean with the nutrients found in colder depths, perhaps by inducing the flow with warm water from a power station or from a nuclear reactor itself, or by bringing the cold water to the surface through a long tube. Natural upsurges of this type, including the Peru or Humboldt current off Peru and the Benguela current off the west coast of Africa, are often biologically productive. However, the problems involved in these and other approaches are so immense that for the moment the sole practicable method of boosting world protein supplies from the oceans appears to be through conventional fishing.

11 · Winning More Food for the World

The Threat Of Hunger

There have been frequent forecasts of global famine in the last two hundred years but happily they have in the past been proved wrong. However, that is no guarantee for the future. Already two-thirds of the world's population—some 2·5 thousand million people, representing almost all of Africa, Asia, and South America—are underfed. Half of these suffer from protein malnutrition. Many millions may actually be starving: no one can actually find out how many. Moreover, the world population is rising rapidly. Before the next generation grows up, the population, now around 3600 million, will double; and then it will double again by the year 2040, while our children are still active. The population increase will be much more substantial in the developing countries, whose people are already hungry (see page 274).

Imbalance between different countries is the worst aspect of the problem. While many of the industrial countries have applied mechanized farming to such good effect that they regularly have surpluses to dispose of, developing countries by and large cannot grow enough to feed their burgeoning populations, and have to buy from the industrial world food they cannot afford, sinking themselves in debt. Their plight is worsening. Because their peoples will increase much faster than those of the industrial countries, 85 out of every 100 additional people between 1965 and 1985 will be in the already poor countries, many of which are already burdened by the size of their populations. This growth of population alone would require an 80 per cent increase in food supplies by 1985 compared with 1962, even without any improvement in the quantity or quality of individual diets.

Yet quality is also important. The world's staple diet is supplied by four chief crops, maize, wheat, rice and potatoes, but these contain only 'low-quality' proteins that lack the right balance of essential amino acids for Man's use. In Britain and other highly developed countries these vegetable proteins can be supplemented by 'high-quality' animal proteins with a better amino acid balance and on average each person manages to get about 60 g (2 oz) of animal protein a day.

By contrast, many millions of people in the developing countries, notably the Far East, are lucky if they get 7 g (0·2 oz) a day. Lactating mothers and newly weaned children are worst affected by the lack of good protein, but malnutrition saps the strength of all the people, making them listless and dull, and unwilling to help themselves. The protein deficiency runs into millions of tons annually, and there is an urgent need for immediate supplies of cheap protein in large quantities to meet immediate needs—protein in a form acceptable to the local people, that can be supplied in a way that would not load the developing countries with a crippling debt for many years. It is proving a major task in those countries, where typically the land must support well over half the population compared with 10 per cent or less in the industrial countries, and where the annual income it brings is a mere $120 per person compared with $3000 in the United States.

The dissimilarity in expectations and living standards that technology has brought to different lands is nowhere plainer than in the food that different peoples have to eat. In the developed lands, about a third of the diet is in the form of animal products; in the developing countries, 90 per cent or more is in vegetable products. Thus the average seventy-year-old North American will have consumed during his lifetime, 2400 chickens, 150 bulls, 20 pigs, 20 hectares (50 acres) of fruit and vegetables, 10 hectares (25 acres) of cereals, and 26,500 litres (7000 U.S. gallons) of water. And the richer countries' food consumption patterns are changing towards better diets. As people become wealthier they eat less bread and turn to more protein-rich food. At one time the staple English diet was bread, tea, jam, butter, sugar and potatoes, with meat as an occasional treat. Now people are spending more on cheese, eggs and meat, and less on bread. Fruit is being diverted from jam making towards freezing and canning.

Meanwhile, people in the developing countries have to eke out their meagre diet of rice, or cassava, say, with meat or fish from time to time. In the past, developing countries have managed to raise production by taking more land for farming—often from pasture and forest—while yields on a given area have remained static. Probably just under half of the world's cultivable land is actually cultivated, but the proportion is slowly rising. Supplies of land for agriculture, however, are limited. Although only about 10 per cent of the world's land surface is farmed, large areas of the remainder are too poor, too dry, too cold or too hot to be cultivated. Within the densely populated areas, and especially in the industrial countries, many users compete for land. In Britain, for instance, agriculture has been losing something like 20,000 hectares (50,000 acres) of land each year to towns, reservoirs, motorways. Every 1·6 km (1 mile) of dual three-lane highway takes 10 hectares (25 acres) of land. Many other countries, particularly in Western Europe and Japan, are in a similar plight. As populations grow, the pressure will become stronger, and farming will have to be conducted on less and less fertile land.

There are naturally ways in which technology can assist in claiming new areas, e.g. the use of irrigation. It is estimated that the 14 per cent of the world's crop-producing land under irrigation grows 25 per cent or more of the total crop harvest. In many countries irrigation offers the only plausible way of increasing food production, as was realized by the ancients, and the amount of irrigated land is expected to increase by 50 to 100 per cent in the next twenty-five years. Again, arid land can be reclaimed: an oil-based 'mulch' can be sprayed on drifting desert sand dunes, stabilizing them and sealing in moisture that would otherwise evaporate. This permits plants to grow in the desert, or heavier yields to be grown on existing arable land. Nonetheless, there are snags. Desert irrigation schemes are almost certain to become focal points for bilharzia, malaria and other diseases transmitted by insect or other forms of life. Over-irrigation without proper drainage can make the ground waterlogged and salty. Besides, the crops produced are particularly vulnerable to attack by locusts or other pests, since these may be provided all the year round with the right breeding conditions. In any case, irrigation by itself will not be more than a palliative. Technology must also be applied to raise the productivity per unit of cultivated land.

Basically, these techniques amount to providing the plants or animals with the right conditions in which to grow—giving them enough food and water, light and air—and protecting them from their natural enemies. The technology is ready, as we shall see in a moment. In the industrial countries modern, intensive, mechanized farming methods have increased yields substantially.

Yet although the technology exists to create mass-plenty for the mass society, by itself it is not enough. In South America, most African countries, India, the Middle East and Asia hardly more than a handful of farmers have any knowledge or understanding of production methods that are commonplace in the industrial countries.

First, millions have to be weaned away from their centuries-old methods of husbandry and taught how to farm more productively, without at the same time wrecking the delicately-balanced environment (see page 263). Bad husbandry and overcropping may destroy the soil's natural structure so that it crumbles into dust—as in the great dust bowl of the mid-United States. In Africa, desert is continually encroaching on steppe, steppe is invading savanna, and savanna forest, primarily because of overgrazing, cereal cultivation and the felling of trees for firewood and charcoal, much of it undertaken within living memory.

Nowhere else can technical capability be so hamstrung by mass inertia, ignorance or poverty as in the fields. The fundamental barrier to change, as always, is human resistance. New types of food, like new ways of farming, cannot be forced down the throats of people bound by centuries of traditional practices, religious and social customs, taboos and other constraints. Those who wish to teach different methods of cultivation—notably United Nations field workers—find themselves involved in changing the social system under which the cultivation is carried out. In South America, for instance, technical change is held up because land is in the hands of a small number of people who, for reasons that are as much psychological as anything else, are opposed to it. Those administering aid schemes, the means by which these new farming practices are introduced, will look for progressive farmers to carry them out. But the progressive ones are more likely to be city dwellers who buy land. By so doing they may dispossess peasants, and so stir up bitterness. In any case, the introduction of machinery displaces the peasants from their customary role, if not from their work entirely. It can create animosity and resentment. In countries where a crowded population is striving to eke out a living on a small supply of arable land, as in the developing countries, an explosive situation can arise. In the Far East, since the new agriculture requires that water must be on hand when the farmer needs it, and not only when the monsoon comes, reliable irrigation is needed. The new methods—which involve growing high-yielding varieties of rice, say, with extensive use of fertilizers, pesticides and herbicides—can only be applied by the best farmers who alone can pay for the expensive materials and the higher labour costs of intensified cropping. Their land holdings are the only ones large enough to make pilot programmes work. So the big farmers succeed at the expense of the small, who flee to the already crowded cities in search of better opportunities. Meanwhile, the arrangements for storage, transport, distribution, marketing prove inadequate to bring food to the places where it is needed at the right time. Those who wish to introduce new techniques to agriculture find that they also have to change social customs and introduce new economic practices, new methods in industry, transport and services before new agricultural methods can take hold (see page 326).

Nevertheless, recent increases in the harvests of wheat and rice in the developing countries have raised hopes that food production can in fact be boosted substantially in spite of the difficulties. Overall world food production rose by 3 per cent in 1967 compared with 1966. In 1968 it rose again, and the 3 per cent increase in the developing countries was actually at a higher rate than the population growth (2·6 per cent). In 1969, however, world farm output remained almost static, and although output rose in developing countries, population increased more rapidly so that output per person fell slightly. It is therefore too early to see whether the application of technology to agriculture is paying off.

According to pessimists, the battle to feed humanity has been lost already, and before the year 2000 hundreds of millions of people will starve to death unless plague, nuclear war, or some other agent kills them first. On the other hand, optimists believe that the 'green revolution' in the fields is just starting, and that the seeds of future plenty have already been sown. Only time will tell which opinion is correct.

Mass-production Farming

Man has always needed the fruits of the earth, but never in such prodigious quantities as he does now. In order to increase the yields from the land in terms of crops and animals, and to safeguard them after harvest, technology has been called upon to impose mass-production methods upon the craft skill of the primitive farmer; to transform methods of husbandry evolved for one man into methods suitable for machines that enable one man to do the work of hundreds; and to sow and reap crops for thousands. To grow healthier, more abundant harvests, Man has to alter the surroundings in which plants and animals grow, protecting them from their natural enemies and providing them with the proper nutrients, warmth, water, and so on. That is, he has to create a more favourable environment for them, to breed them in larger numbers than could normally survive in the natural state. This he can do in a variety of ways. Though the principles of growing crops have been appreciated to some degree since Man the hunter settled down to till the soil in Neolithic times, intensive methods of agriculture have only become possible in modern times through applying modern technology.

Technology started to overtake agricultural practice in the nineteenth century when the introduction into western fields of the seed drill to sow seed, of harvesting machines—first for grain and then for other crops—and of various hay-making machines, made farming less dependent on human muscle. Horse-drawn cultivators and harrows became common for preparing the land, while iron and steel were fashioned into ploughs and other implements. The first tractors driven by internal-combustion engines appeared on farms around the beginning of the twentieth century, providing a multi-purpose farm tool that eventually displaced the horse as the prime mover in the industrialized world. Fields have been steadily enlarged, to give machines clear runs of as much as several kilometres at one go—sometimes even further—and this is having a strong influence on the appearance of many a countryside. Elsewhere, however, animals such as oxen, elephants, water-buffalo, horses, donkeys are still harnessed to pull farm implements, carry loads, thresh grain, raise water, and perform other tasks.

All these inventions paved the way for farming almost exclusively with machines on mass-production principles. Crop production in the industrial countries is now highly mechanized, and such aids are being used on an increasing scale in the rearing of livestock and ancillary farm tasks. There are now few jobs that cannot be mechanized provided there is the money to buy or hire the equipment, and the skilled men to handle it. The past thirty years or so have witnessed the introduction of machinery for the bulk handling of grain, animal feed and fertilizers, and of irrigation apparatus for spreading water or liquid manure on a large scale.

Indeed, here in the fields it is possible to see the vast disparity in technological levels between various communities (see p. 37). The plough is agriculture's basic tool. The poor peasant farmer toiling in the dusty fields with a wooden plough whose design has not changed for 5000 years or more represents a low level of technology. At the other end of the technological spectrum are farmers using steel multi-furrow ploughs, with automatic depth control, drawn by a tractor with the power of more than a hundred horses, enabling one man to plough 6 hectares (15 acres) in a working day with little effort. Other types of plough using rollers or ultrasonic waves to cut the ground are being investigated. A more radical type of agriculture is in the offing which may transform agriculture again in the future as much as mechanization has done. In this, drills inject ammonia or other nitrogen-containing nutrient chemicals several centimetres down through a small hole without disturbing the bulk of the soil. The plough itself may have been superseded, in some regions at least, by chemical sprays of contact herbicides such as diquat and paraquat that kill off old growth from the previous season and allow seeds to sprout through the dead herbiage. By then adding pesticide to the liquid fertilizer and spreading both at once, the farmer can halve the time and labour involved.

Harvesting is one of the most difficult operations to mechanize, but technical ingenuity is paying off, and there are now many kinds of machine that can harvest crops as diverse as rice and oranges. One of the most familiar to the western world must be the combine harvester, which propels itself through fields of cereals, cutting them, threshing to get the grain out of the straw, and dropping the chaff behind. One man driving such a machine can reap in a day as much grain as 300 men with sickles. What is more, the machine can work on after dark, lighting the field with spotlights (Diagram 11.1). Potatoes can be gathered by a tractor-drawn machine that lifts and grades them, and then loads them into wagons drawn alongside by

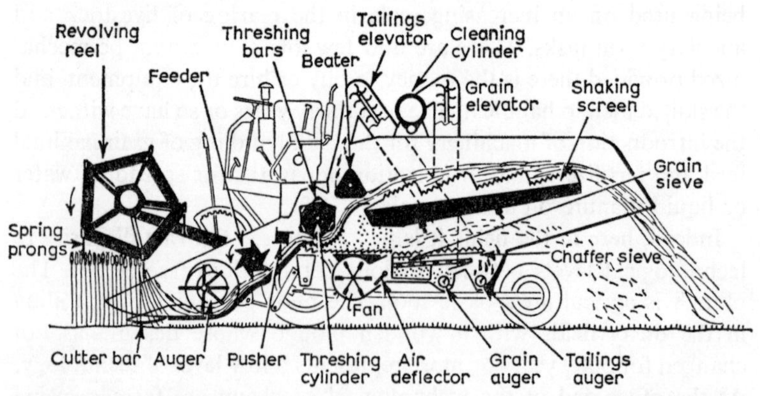

Diagram 11.1 Principle of the combine harvester, which cuts and threshes grain in one pass

another tractor. Machines for 'picking' cotton by stripping it from the stems can gather a bale of cotton in thirty minutes, whereas a man would take forty to fifty hours picking it by hand.

Once the crop is gathered in, grain silos and grain-drying machinery, for example, handle hundreds of tons an hour so that the crop can be preserved in its best condition. Then the harvest can be safeguarded by the application of chemicals, or by irradiation, in order to kill pests and prevent micro-organisms gaining a hold. This helps to cut down losses during storage—often a major factor. It is estimated that one-quarter of the world's output of crops is eaten or destroyed by pests, moulds, etc. The preserving of foods by radiation is quite a new technique, and much investigation is going on into the safety of the practice, but already certain irradiated products—mushrooms, for instance—have been sold in shops in Europe and elsewhere.

Besides tilling the soil, Man can improve the chances of his crops by adding nutrients to the soil where these are lacking, or irrigating where the land is dry. Since the discovery of artificial fertilizers in the 1840s, they have been applied to the land in increasing amounts, replacing natural fertilizers to a large extent in many parts of the world. In many advanced countries—and to an increasing extent developing countries also—fertilizer manufacture is a major industrial activity, and a large consumer of raw materials, chiefly those containing nitrogen (which is converted to such compounds as ammonium nitrate), phosphorus (as phosphates), potassium (as potash) and calcium (lime). These it obtains from one part of the environment; it then processes them and distributes them to others.

A third of the world fertilizer production consists of nitrogen compounds. Most of them are derived from synthetic ammonia, which is itself made chiefly from hydrogen in oil and nitrogen from the air (see page 72). The graph shows the growth of fertilizer consumption over the last twenty-five years or so (Diagram 11.2). However, farming methods differ markedly across the world, and at present the developing countries take only 12 per cent of total fertilizer production—although they have two-thirds of the population—largely because they cannot afford more, and do not yet practise enough of the sort of farming that uses synthetic fertilizers.

Plants have natural enemies—other plants that choke them, insects that prey on them, diseases that infect them—and Man can protect them by applying agricultural chemicals, e.g. selective weedkillers

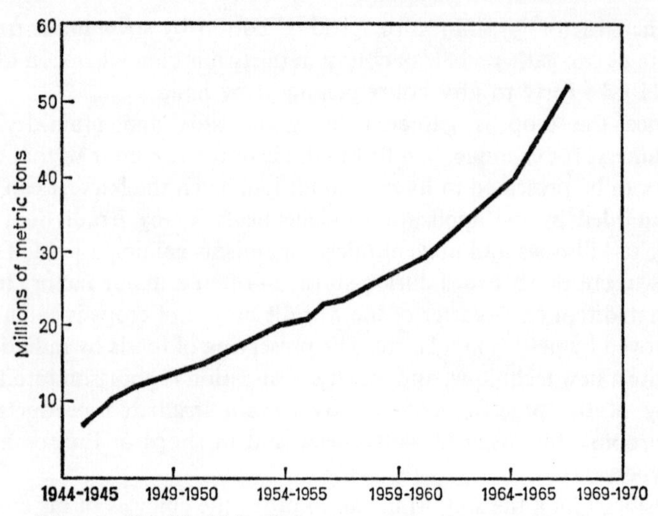

Diagram 11.2 World fertilizer consumption in recent times (UN Food and Agriculture Organization)

and hormones that control plant growth. Though compounds of arsenic and copper have been used to destroy insect pests since about 1870, the development of spraying and dusting machinery for applying them on a large scale has taken place since about 1942, after the first major synthetic compound, DDT, was used extensively by the Allied forces to rid war zones of lice, mosquitoes, beetles and other unwanted life. Sprays to keep off insects and other pests can now be applied to the crops by aircraft or tractor-mounted boom sprayers, with which one man can treat as much as 60 hectares (150 acres) a day. Less can be done about locusts and plague insects that descend on crops and strip them bare in a few hours, though airborne or radar tracking, and bomb or chemical attacks from the air, provide control of a kind.

However, the intensive cultivation practised in the industrial world does provide great areas of one type of plant which represent an ideal breeding or feeding ground for the plant's predators, and for disease. Another drawback is that the over-enthusiastic application of agricultural chemicals has in certain areas such as the United States been reflected in pesticide residues being concentrated in the bodies of birds and animals, and hence in Man. The effects have been considered so serious that in some cases use of the chemical concerned has been banned, or else severely restricted. Certain residues now appear to have been carried into all parts of the world (see page 263).

Sometimes, farmers can adapt or secure the microclimate in which plants are reared by growing—or building—barriers against harmful winds or against cold, or by enclosing the plants inside greenhouses, giving them more heat and light than they would normally receive. Where he cannot ensure the proper microclimate, Man may be able to breed new strains of plants that are hardier, more prolific or improved in some other way to find those that will thrive in previously unfavourable conditions. For instance, 'miracle rice', a cross between a tall Indonesian strain and a short-stalked breed from Taiwan, matures rapidly, yielding two or three crops a year (see page 218). Various methods can be adopted to produce these breeds, such as deliberately cross-pollinating strains, and irradiating seeds or plants. By these means the productivity of wheat, for instance, has been raised so that some varieties used in progressive agricultural countries yield almost ten times the world average for a given area.

Nevertheless, the value of being able to grow a certain crop in a certain place is offset by the cost of doing so. If it would theoretically be possible to grow potatoes at the South Pole, the cost would put it out of reasonable bounds. To give a less extreme example, it is more costly to build, maintain and heat a glasshouse than to grow crops out in an open field, but in a hostile climate it may be thought the best alternative to not having the crops at all.

The rearing of animals, too, is being put on the same mechanized footing as the production of goods in the factory, giving rise to intensive rearing—particularly of hens, pigs, and cattle, many of which spend their lives indoors and only emerge into the open on the way to market. In the United States, over 90 per cent of all chickens, 30 per cent of turkeys and nearly 40 per cent of pigs are born and bred under controlled conditions. The animals are regarded as units for converting given amounts of feed into outputs of protein, fat and so on, for human consumption. In order to minimize the space they occupy and make the farmer's job easier, they are often enclosed in small pens which allow them enough room to move but not enough to dissipate their energies in free movement. Feed and water are supplied from hoppers through chutes or pipes at rates that are carefully controlled so as to make the animal grow, or the hen lay eggs, at the optimum rate. Such methods can bring quicker, fatter returns than free-run open-air farming.

To economize further, multi-storey pens can be constructed. However, vertical cultivation, whether of plants or animals, tends to create more problems than it solves. Some experiments have been carried out with growing plants (e.g. tomatoes) in pots fixed to a vertical travelling band inside a glasshouse tower; but they are unlikely to be widely adopted. For the most part, agriculture and horticulture remain firmly committed to horizontal practice.

However that may be, the method of 'factory farming' that is now common throughout most of the industrial world has been condemned as inhuman and cruel, even though it makes economic sense. It is of course completely unnatural and brings some tangible evils. Lack of exercise and sunlight weakens the animals. The inability to indulge in their normal activities—rooting, scratching, etc.—makes them bored and leads to distress and attacks on other animals. Disease can spread rapidly within the building in which the animals are kept.

New Methods in the Food-supply Chain

The pressures to grow more food are paralleled by pressures to process and preserve food more quickly through factories, with fewer employees and at greater speed. The origins of these pressures are to be found—like so many of the influences on our society—in the Industrial Revolution, which changed the structure of society and concentrated large numbers of people in towns and cities, altering habits and creating massive markets for products in large quantities and of identical quality.

These pressures have prompted the application of mass-production methods to foodstuffs (in a way similar to the processing of materials, see page 109), as well as mechanization on a considerable scale in the preparation of food, and in the chain by which food is brought from farm to consumer. In the highly industrialized countries few of the consumers are farmers, and so food has to be brought to them by a long chain of suppliers, transport operators and middlemen.

For instance, in the industrial countries daily bread is not touched by hand at all. It proceeds automatically from the weighing out and mixing of flour, water, yeast and salt to the time that the wrapped, sliced loaf is dispatched for sale. Vast continuous ovens (Plate 11.1) turn out an endless stream of loaves all cooked to exactly the same standard brown, and all as nearly reproducible as it is possible for technology to make them. The public benefit by the higher, more consistent quality, and better choice of more exotic foods: but standardization is the norm and variety commands a high price.

Two of the most pronounced trends in food preparation and distribution are the growth of the market in convenience foods—where a high degree of preparation is done by the supplier—and the increasing concentration of trade in the hands of mass suppliers, both producers of goods and the supermarkets that sell them to the public. These trends have been accentuated by such technical aids as canning, deep freezing and dehydration, as well as the acceleration of transport, all of which demand heavy capital investment, long production runs and hence co-operation between large suppliers and wholesalers.

For centuries, Man had no better way of preserving food than drying or using salt or vinegar. Then canning was invented in the nineteenth century, when it was found that food sterilized by boiling could be preserved in canisters. Napoleon's forces ate food that had been sterilized and sealed in glass jars. Britain capped that by using tinplate canisters. Initially canned foods were chiefly used on sea voyages or journeys of exploration, and as military rations. The cans were made, and the lids soldered on, by hand. But by the early twentieth century automatic or semi-automatic machines had brought costs down, and made them commonplace. Now canneries are extensively mechanized and canning is a mass-production, mass-consumption business. Tasks like that of sorting out discoloured beans or peas, carried out by groups of dexterous mob-capped women as recently as the 1930s, are now done by electronic sorting apparatus or other machines.

The freezing industry is also big business. The process has been made easier since ice-making machines of the later nineteenth century, most of them working on the cooling effect of the evaporation of liquid ammonia or other volatile liquids, were supplemented by very cold liquefied gases, notably nitrogen made in bulk by cryogenic methods (see page 74). Accelerated freeze-drying (AFD) became feasible with efficient pumps that could extract water vapour rapidly under vacuum from the freezing food. The AFD process is said to leave protein and vitamin in the food unchanged, and taste little affected. Treated foods can be stored in small spaces and rapidly reconstituted with water. It has succeeded with certain brands of instant coffee, for example, but production costs are high. The freezing of foods using liquid nitrogen (at −196°C) is quicker than conventional methods of cooling either in a cold air blast or on a refrigerated plate. Modern freezing units can handle astonishing amounts of products, ranging from whole fish to apricots, and poultry to bread rolls, in a very short time. A fully mechanized production unit in which food is precooled by nitrogen vapour and then frozen stiff by a nitrogen spray, can be controlled by one man. Large-scale refrigerated storage and transport in ships, trains, lorries and even aircraft is common. Foodstuffs can be traded with almost as much ease as minerals or manufactured goods, allowing a temperate country to supplement its diet with produce from tropical countries, and vice versa.

Mass production is turning the refrigerator from an item of domestic affluence to a felt necessity, and one which is helping to revolutionize eating habits. Frozen foods now embrace such stalwarts of the western diet as fish, peas, beans and meat products, but there are many other items from cream cakes to mousse. In Britain the habit of buying frozen food is not so compulsive as in certain other lands: only 2 per cent of all spending on food goes on frozen foods and the average annual consumption is 5 kg (11 lb) per person. In Sweden, though, it is about double that, and in the United States about six times as high. Manufacturers look forward to the day when more British households have an adequate refrigerator: only about 50 per cent do at present, and only about 2 per cent have freezers, which are needed to store frozen foods in any large quantities. At least more shops are installing bigger and better freezing display cabinets.

Alongside the adoption of new food processing methods have been developments of new packaging techniques without which the 'convenience' aspect would be impossible. What began long ago as the simple action of wrapping up produce in leaves and carrying wine and water in goatskins has been transformed into a whole industry devoted to making the product more attractive and durable, protecting the product and persuading the customer to buy (but see page 271). Plastics have had a major impact, especially in the lines of transparent film packs that protect the food while letting customers assess the contents. Potato crisps, for instance, used to be sold in bleached greaseproof paper packs, and had a shelf life of seven days. When cellulose film came in for packaging, the shelf life was increased to four weeks. Now, swathed in polypropylene, the crisps last six weeks. With the steady increase in self-service retailing during recent years there has been a tremendous growth in the number and variety of pre-packaged goods including foods, clothing, hardware, cosmetics, motor accessories, confectionery and many more.

At present, nearly a quarter of the weekly food bill in the average British household goes on convenience foods—instant coffee, tea, milk, potatoes, and even 'instant meals', too—although they seem to be more expensive sources of energy and vitamins than ordinary foods. Convenience foods have until now been designed as parts of the ordinary meal. However, they could become meals in their own right for busy people. This sort of thinking is already being applied in industrial canteens. Part-cooked portions are placed in aluminium foil trays and sealed with lids. Then the trays are cooled in a blast-freezing tunnel and transferred to a central cold store and left until needed. They are sent to local catering units (in a factory canteen, perhaps) in insulated or refrigerated containers, and put in the unit's cold store until needed. When wanted, their cooking is completed by electric or microwave ovens.

But the meal need not be frozen. An entire pre-packed meal could be served up in tins or packets, or squeezed forth from a tube like toothpaste. 'Tube foods' and dehydrated foods—entire meals reduced to tablets, cubes and pastes, devised for American astronauts to eat while in space—demonstrate that it can be done. However, the eating of convenience foods means that no longer need the whole family be involved in preparing and eating meals together. It is easier for each person to get his own. The Sunday meal around the joint and vegetables is no more the focal point of the week, just as going to church is no longer its hors d'oeuvre. The tendency is to eat more snacks in place of fewer set meals and miss out breakfast. One meal in three in the United States is eaten away from home. Forecasts suggest that it could be one in two by 1980. Food processing can produce foods that are both cheap and nutritious, and also attractive, tasty foods of little nutritive value, in large volumes. It is easier to choose a bad diet than it was fifty years ago: will it be easier still in twenty years' time?

There have been changes in the distribution of these and other goods to the public, with emphasis since the mid 1950s on self-selection and self-service, and the growth of multiple chains. These in turn make manifest changes in technology and method—the trend towards more expensive buildings and plant, less labour and lower stocks, the employment of modern management methods to distribution and the realization of economies of scale. Most significant has been the rise of the supermarket, with its convenient layout of many kinds of item under one roof, the benefits of self-service for the customer, and reduced labour costs for the operator. In the United States, 90 per cent of food is sold in supermarkets. In Britain, there were an estimated 2800 supermarkets in 1967, holding 21 per cent of the grocery trade; by 1977 this number is expected to have increased to 7200, with 60 per cent of the trade. The supermarkets are also expanding in size and are carrying a greater range of goods.

Slowly but surely, supermarkets are becoming the self-service department stores of the 1970s. With greater mobility of shoppers, many of whom have their own cars today, shopping centres are bound to become more popular. Hence the 'convenience' stores sited outside towns and having a complete range of food shops, chemist, drycleaner and post office will be where the housewife drives once or twice a week—with the family on Saturday. Other groups of shops in the town centre selling clothing, furniture, shoes and other consumer goods will remain, although it will be easier for the out-of-town shops to lay on the lavish parking space needed.

Fortifying Foods

There are three chief ways to fortify weak diets with protein. One is to improve the quality of the nutrients in the crops on which much of the world depends—grains such as wheat, rice, maize and barley. Another is to add protein concentrates to these natural products. The third is to add not whole protein but separate amino acids to correct a deficiency of one or more of them. Man cannot use all the amino acids present unless there is a certain minimum of the others also present (see page 228); for example, some of the amino acids in maize products are wasted unless there is more lysine there than occurs naturally. By careful genetic selection it is possible to produce strains containing naturally more lysine and more overall protein than the parent strains. This has been done with maize, for instance. The great advantage is that the extra goodness is in the grain and nothing has to be added afterwards.

The addition of protein concentrate after harvesting is more awkward, though it can be done. The concentrated protein can be extracted from fish, leaves, or seaweed, or single-celled organisms. However, these concentrates may change the taste, texture and cooking properties of the food—not necessarily in a desirable way. Fish meal was originally used as a fertilizer, and then as an animal feed. By a refinement of the original method of cooking, defatting and drying, the fish flesh can be converted into a form suitable for human consumption. More recently, fish protein concentrate has been made in the United States and elsewhere from the whole fish—bones, viscera and all—which is cooked, defatted, dried and ground to produce an odourless powder, finer than fish meal.

Leaf protein concentrate, as either a tasteless liquid produced when leaves are pulped, or a pressed cake made by coagulating the juice, is the cheapest concentrate at the moment. It can be fed to pigs or other animals directly in a slurry feed. Oilseed meals—left after the oil has been extracted from seeds like cotton, soybean, sunflower and peanut—contain about 50 per cent protein which can be got at low cost. An enriched mixture of one-third oilseed meal and two-thirds cereal grain contains more than 25 per cent of relatively good-quality protein. This is the base for many cheap, high-protein vegetable mixtures, such as flour and soft drinks.

Another possibility is to extract the protein from one foodstuff (oil seeds or similar produce), concentrate it in almost pure form as lumps, flakes or fibres, and 'retexture' them: the protein can thus be spun into fine threads in the same way that synthetic textile fibres are made. The threads can be compacted into blocks, with appropriate colour and flavours. It is said to be difficult to tell the synthetic product from the real thing, be it bacon, beef or chicken, and these products are enjoying some commercial success in the United States and other places.

Again, single-cell protein can be extracted from bacteria, yeasts or fungi grown in certain petroleum products (oil or natural gas forming the basic material), or possibly vegetable starch products. Most of the major oil companies are studying the production of protein from petroleum sources. If the processes turn out to be economic, they could provide large quantities of food, albeit initially for animals. Current world crude oil production is around

2000 million tons a year. Using certain fractions of this it would theoretically be possible to produce 20 million tons a year of protein concentrate, containing 65 per cent or more of protein by weight; but since not all refineries would be capable of producing it, a more realistic figure would be around 10 million tons a year. The method adopted by British Petroleum, one of the furthest advanced at present, is typical of the general concept. Certain strains of yeast are grown on an oil-water mixture, being fed with sources of oxygen, nitrogen and mineral salts. Then the yeast is filtered off and dried. There are actually two routes, which start with different oil fractions from the refinery. They are illustrated in the Diagram 11.3.

In both cases the initial target is to make protein that can be fed to animals—chiefly pigs and poultry. The next stage is to try it on humans.

Such biological treatments can produce edible protein from the most unpromising sources, including bagasse (the fibrous residue left after sugar is extracted from sugar cane), old newspapers and even coal. Strict testing on animals for toxicity, wholesomeness, etc. must of course precede any addition of the products to human diets. Yet another source is algae, which, in contrast to most other growths, are able to draw their energy from sunlight and do not need to be provided with an energy-rich 'broth' in which to grow. Another point in their favour is that some tribes living on Lake Chad in the heart of Africa have collected and eaten algae since ancient times. At least the kinds that they gather are palatable.

The modern protein-producing processes, intriguing though they are, stand or fall on costs rather than on technical ingenuity. They are not feasible on a large scale if the foodstuffs they produce are more expensive than conventionally grown supplies. Also, looked at through the nutritionist's eyes, it is much more efficient to feed humans directly than to feed animals with the protein, and then eat the animals. The exception is when the food is not a protein but grass, hay and similar materials which the simple human stomach cannot digest. Many new vegetable drinks and gruels, enriched with protein, are on sale in developing countries, and some are well established and commercially successful. Soy milk for feeding babies is now accepted, and 'filled' and artificial milk are on sale in the United States and India. By no means all of these prepared foods, however, have been accepted by the people they are supposed to nourish. Certainly, it can often take a great deal of care to prepare synthetic or fortified foods in an appetizing form, and several products have failed to attract their intended consumers.

Amino acids, and vitamins, for fortification may be mass-produced by fermentation or chemical synthesis and do not affect the taste or other properties of foods. At present amino acids cost something like $2 to $50 per kilogramme, but these prices should fall as demand and production increase. They could be added to a great

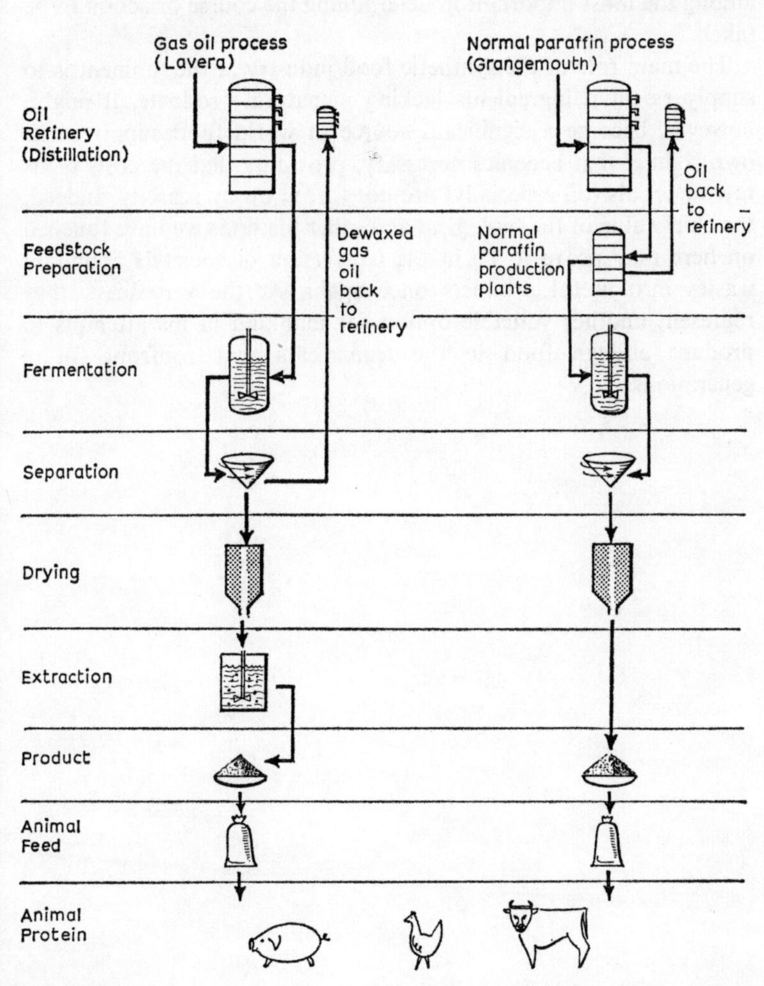

Diagram 11.3 Processes for the production of BP protein concentrate—'food from oil'

variety of foods, and so the choice can be made freely according to each country's needs. A programme of lysine fortification of flour in India, for example, is estimated to cost $1 per person per year for one million people. Bread with added lysine has been baked there since the beginning of 1969. It is often difficult to find a suitable way of administering these concentrates: it may be done centrally, in large mills or in small mills in the villages. Again, social factors are among the most important in determining the course of action to be taken.

The main role of the synthetic food industry at the moment is to supply essential ingredients lacking in natural products. It might, however, become a significant source of world food supply in its own right if that becomes necessary, providing that the costs of its raw materials (oil especially) are not forced up by scarcity. Indeed, the chief value of the biological and other methods we have touched on here may prove to be in the conversion of society's abundant wastes into useful products once again. At the very least, they represent another valuable option for mankind in his attempts to produce enough food in the lean years that confront future generations.

12 · Construction and the Environment

The Engulfing City

Few of Man's activities have as much effect on his surroundings as his building and the excavations he makes to win building materials. The built environment is the embodiment of human will dominating Nature, and a constant victory over the levelling force of gravity. Visibly, Man's works are swallowing up the landscape as he drives forward metalled roads and railways to carry him and his goods rapidly across difficult terrain; spans rivers and chasms with bridges; and constructs buildings that surround him in a comfortable micro-environment protected from the elements.

All this ramified technical infrastructure is hewed out of, and grafted on to, the natural environment. In time it comes to swamp it. In some city areas not even an occasional tree breaks the barren monotony: nothing is there to remind the inhabitants that there was ever anything but steel, concrete, glass.

Yet while the scope of Man's domination has widened substantially throughout history, notably since the Industrial Revolution, the pattern of his settlements has altered remarkably little, and noticeably so when it comes to the basic unit of human occupation—the dwelling and its plot of land. From an aircraft flying a few thousand metres above the earth, it appears as though technology had hardly affected Man's pattern of living. Such clusters of little boxlike houses, arranged around roadways, were familiar to the Babylonians. The method of putting them together from thousands of small bricks was well known in ancient Egypt.

Man's basic needs of food, shelter and warmth—and indeed, Man's instincts—have hardly changed at all in the 6000 years or so since he started to build. Because of this, ideas of what constitute a desirable house are very much as they have always been. Most people still live in essentially small, boxlike places, where they can indulge their own tastes and create their 'personal environment'. Elsewhere, a man lives amid surroundings chosen by others. Only within his home can he please himself.

The distinctive aspect of modern society is the scale on which it has grouped these dwellings together, along with other constructions, into huge agglomerations. The city so formed is the home of a growing proportion of the world's people, and the proportion is likely to continue to rise. Urbanization is one mark of an increasingly technological economy, brought about by the demands of mass society. Around 10,000 years ago there were not more than about 10 million people on earth, about as many as now live in greater London, and the largest settlement consisted of around 8000 people. The city as we would recognize it probably traces its origins back to between 5000 B.C. and 2000 B.C., in southern Mesopotamia, the land of the Tigris and Euphrates rivers, and the fertile crescent where writing was invented. By the year 2000 B.C. the total population had risen to perhaps 100 million and the largest city held 100,000 people.

These early cities were small islands of man-made order, subdued out of the wilderness by primitive technology. When their technology proved too weak to control their surroundings, they vanished.

But many cities grew and prospered. People flocked into them in search of a richer life and better opportunities. The process accelerated markedly with the Industrial Revolution. Before 1850 no society could be described as predominantly urbanized. By 1900 only Britain was in that condition. Yet now all industrial nations are highly urbanized, and the developing nations are rapidly following the same path.

The modern growth of cities owes much to technological services that allow people in different parts of the city to keep in contact with each other, and can supply the city with its nourishment in the shape of food and water and remove the waste products, sewage and rubbish. Until the nineteenth century these factors prevented massive expansion. But with that century, factories appeared, providing large numbers of jobs in concentrated areas. Then came railways, which enabled people to travel easily and more quickly from one part of the city to another, and from the city to the surroundings, and made it possible for fresh food to be brought in from a large area round about to meet the needs of the working population. Products, too, could be taken out more easily. Tramcars, underground railways and motor vehicles later made these transfers more speedy, allowing cities to grow larger.

The disposal of sewage and other wastes is always a problem for densely populated areas, as the cities found out. Two hundred years ago sewage disposal systems were virtually non-existent and the ordinary house, even in industrial Britain, was lucky to have running water. As late as the middle of the nineteenth century, pigs were brought into Manchester city to eat rubbish thrown into the streets. Only one hundred years ago thousands of houses in London had no drainage, the majority of them had stinking, overflowing cesspools, and the city was still in the throes of installing its first comprehensive system of main drains and sewers. Even now cholera, transmitted by polluted water, is a constant danger, unless the water supply is safeguarded, from sewage leaks especially (see page 214). Between 1848 and 1854, a quarter of a million people died of the disease in Britain. Typhoid, too, took its toll. By the late 1860s the scientific knowledge of bacteriology and the nature of infectious diseases was far enough advanced for it to have begun influencing social behaviour. The introduction of cast-iron water pipes instead of bored-out treetrunks or pottery pipes allowed pure water to be distributed uncontaminated by sewage. Later, treatment plants were built to process the sewage before discharge to waterways.

Communications had to be improved if cities were to grow. The electric telegraph and next the telephone were introduced, providing a means of instantly passing messages to and fro across the city. Modern communications, in the shape of radio and television, can now also be 'piped' in. Energy had to be supplied to the home and office. First came gas, then electricity. In major new building schemes there may be district heating. All these services are aimed at making the domestic surroundings more comfortable and render the city a powerful magnet for population and wealth.

In Britain, urban growth, population, wealth and higher prices are becoming concentrated along a line from Manchester through the Midlands to London. The axis continues on to the Continent. Pessimists forecast that soon there will be one huge conurbation stretching from Manchester in the north to Milan in the south. In the United States, urbanization has reached an almost absurd stage. Although it was 1875 before the first 25 per cent of the American population came to be living in cities, by 1960 about two-thirds was urbanized, and by 1970 it was almost three-quarters. By the year 2000, the United States will be a country of 300 million people, and nine out of ten will be living in cities. Half of the people live on 1 per cent of the land, and there is a continual drift to the big cities. In the north-east corridor, 50 million people are squashed into the world's largest urban area, which stretches for 1000 km (600 miles) along the coast from Boston to Washington. In the Soviet Union,

20 per cent of the population lived in cities in 1930, but by 1960 it was 45 per cent. In Japan, where the mountainous nature of the land restricts living space, it is a similar tale. More than 40 per cent of its 100 million people live in 1 per cent of the total land area. Tokyo, which has become the world's largest city, holds about 12 million people within its gigantic sprawl.

In spite of Man's growing technical competence, more than two-thirds of the earth's land surface is sparsely populated—polar and frozen deserts, mountainous regions and hot deserts. Here, too, technology can maintain comfortable living conditions, at a price, but only a very small fraction of the world's population could be supported in such inhospitable areas for an appreciable time. Within the habitable areas, a growing proportion of mankind is crowded into the built-up areas, placing a great strain on the necessary services. In cities of the developing countries, these services are often strained to the limit. Great wealth exists cheek by jowl with abject, unsanitary poverty while the building, health, education and welfare services struggle to keep up. Even when slums are demolished to make way for new high-rise blocks, this brings more people to the area, aggravating the problems of overcrowding, noise, pollution, disease and waste disposal. And still the cities keep growing, swollen by the never-ending flood of humanity.

What will happen as cities continue to grow? The choices, apart from deliberately discouraging city growth—which is not feasible while populations continue to increase—are to let the cities expand sideways, as they always have done, or to build up, or to build underground. All three approaches have been tried. Thus many families rehoused from slums, together with office workers, know what it is like to live or work in tower blocks high above the ground. Some cities have had success with underground shopping centres or car parks. But in the main the city expands outwards, swallowing more land as it does so. In densely-populated areas such as Western Europe the available land is rapidly disappearing, and continual outward growth may soon prove unacceptable, especially if city dwellers are to have any green recreation spaces worth speaking of, close to the city. In any case, there is a real danger that the city will be choked by its own growth. If in the next decade or more the strains will be eased by new transport systems that allow people to live in more pleasant surroundings further from the city centre, there is obviously a limit to the exercise. Looking further ahead, the sole possibilities appear to be to increase the density of population in the existing cities, and to create new ones, either starting on greenfield sites—where these can be found—or taking existing towns or villages as nuclei. The problems of preserving a decent living environment will be among the most acute that the planners of technology will have to face.

New Methods and New Materials

New practices have had to grow up alongside the traditional skills of the bricklayer, carpenter and stonemason. New skills were needed to cater for the arrival of electricity and gas, piped water and sewage, telephones and television, heating and ventilating systems. Each technical invention in its day provided new jobs around the core of the building industry, which today typically employs more than 10 per cent of the population in most countries.

Man's attempts to shape his environment have been radically strengthened by the mechanization of construction, together with improvements in materials and the ways they are used. Thus many powerful machines can be used to aid labourers on construction sites. Huge bulldozers and graders clear the ground; lorries move earth, concrete beams, steel rods and other materials; there are cranes and excavators, concrete mixers and many more. The machines can accomplish herculean tasks, such as uprooting a tree, which would have otherwise been impossible, or required the concerted action of dozens or hundreds of labourers. With their aid, virtually anything that Man wants to do can be done. Nonetheless, amid all this mechanized activity, there are still sites where materials are loaded into a bucket and hauled up by rope and pulley, and men laboriously climb ladders with hods of bricks over one shoulder, just as they have done for centuries.

The Industrial Revolution put paid to the widespread use of timber for building in Western Europe because the wood was wanted for fuel (though naturally those countries that possess ample forests continue to build homes with timber). Bricks and mortar, cast iron, then steel and concrete, replaced it. Cement has been used since antiquity, but until the eighteenth century it was still based on the Roman recipe of burnt lime, sand and water. Not until the early nineteenth century was Portland cement invented—so named because it was supposed to look like Portland stone. This improved cement, incorporated into concrete, has made it possible to build the enormous structures of modern cities, although plain concrete has the basic flaw of being weak under tension. Concrete was reinforced by iron rods, marrying its compressive strength with the tensile strength of the rods. The coming of cheap steel in large quantities around the end of the nineteenth century provided a new structural material, to be used either alone or as reinforcement for concrete instead of iron. As a structural material, reinforced concrete allowed architects to design buildings with much greater freedom. Beams and floor slabs, columns, pipes, bridge supports and many other components could be made out of the new material with great success.

The United States took the lead in producing cheap steel, and also in using it in 'skyscrapers' in the last two decades of the nineteenth century. The evolution of a safe, electric passenger lift at around the same time played a large part in convincing the public to actually work, and live, in these monstrous towers. The next major step came in the mid 1930s with the technique of pre-stressing—by which strong steel cables, passing through ducts in the concrete, are tightened against end plates to hold the concrete in compression. Pre-stressed beams are thin for their length, and so the technique means that houses, hotels, blocks of flats, hospitals, schools and so on can be built with lighter, stronger designs. The quality of the steel around which buildings are constructed has steadily improved, and concrete technology has increased the material's strength by three times over the last thirty years.

Before the coming of steel frames for buildings, the height of stone or brick buildings was limited by the sheer bulk of the materials needed to support the load. Around the turn of the nineteenth century, a building fourteen storeys high had to have walls of masonry that were several metres thick at ground level. The problem was resolved by the relatively light steel frame which could support the loads and on which the rest of the building could be constructed by filling in with relatively light materials, such as brick, aluminium or glass.

As a result, we are now constructing thinner buildings faster and more cheaply than our forefathers did, as shown in Diagram 12.1.

The escape from the principle of having to have a thick, load-bearing wall of brick or stone which insulated the inside of the house

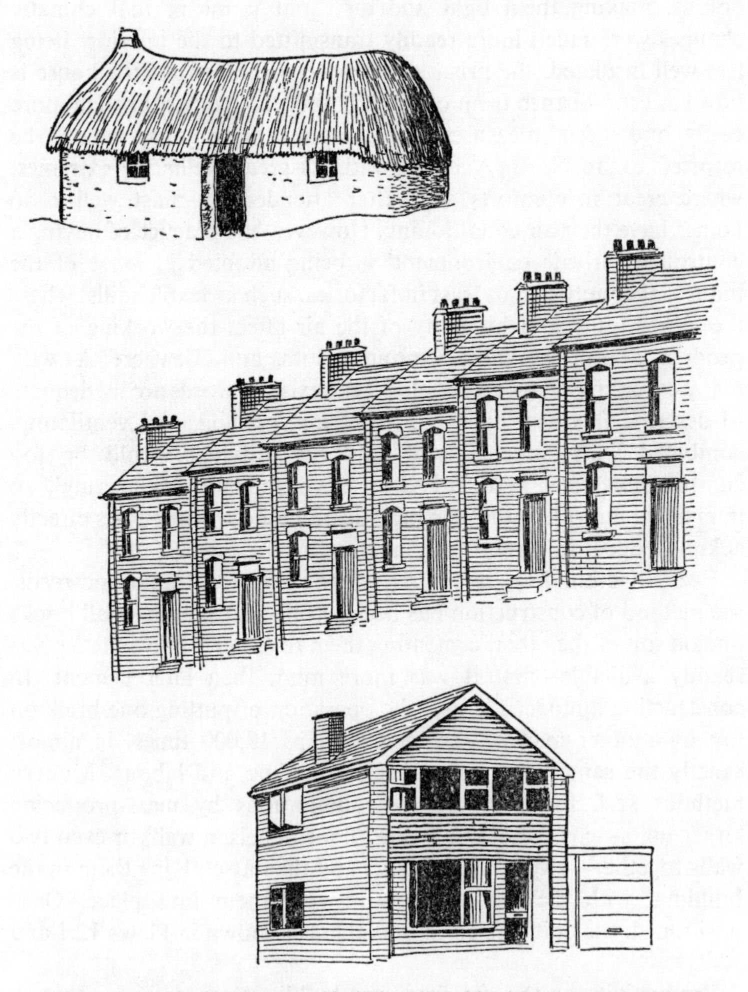

Diagram 12.1 The early type of building, like this rural cottage (*top*) had massive, loadbearing walls of brick or stone, or both, which insulated the interior from the weather. The trend towards thinner walls and larger windows, taken further in the terraced houses of the last century (*middle*)—and very evident in the modern detached house (*bottom*)—has meant that the structure provides less insulation, and thus the conditions inside follow those outside much more closely, unless regulated by some other means

from the elements enabled architects to design large windows in houses, making them light and airy, but it meant that climatic changes were much more readily transmitted to the interior. Being less well insulated, the private thermal environment of the house is now easier to disturb from outside. Noise, too, comes through more easily, and special measures such as double glazing may have to be resorted to. In North America, with its greater climatic extremes, where creature comforts are better attended to, most well-to-do homes have their air conditioning. However, the practice of having a controlled 'private environment' is being adopted in more of the industrial countries, not least in factories, such as textile mills, where the temperature and humidity of the air affect the working of the product, and in hospitals, computer rooms and elsewhere. As with many other branches of technology, the existing words are inadequate to describe this new operation: 'heating, cooling, and ventilating, combined with some filtering and humidifying' would be too cumbersome, so it is described as 'environmental engineering', an interesting indication that the modification of environment is directly acknowledged, in a small way at least.

Ever since buildings were first put together from mud and reeds, the method of construction has been to pile numbers of small bricks one on top of the other, cementing them together with whatever was readily available—first it was more mud, then later cement. In constructing a modern house, the operation of putting one brick on top of another may be repeated perhaps 18,000 times, in almost exactly the same way. This takes much time and labour. Modern methods seek to overcome these limitations by mass-producing larger units—e.g., whole staircases, floor panels, a wall, or even two walls together—away from the site, and then assembling them in the building, with the aid of cranes to lift them into place. Once positioned, the unit is fixed to the others as shown in Plates 12.1 and 12.2.

Productivity on the site rises, and buildings can be erected much more rapidly. Such industrialized methods have reduced the labour content of putting up blocks of flats over four storeys by 30 to 40 per cent, and cut construction time by as much as half. However, the savings made on site are offset by the costly extra equipment needed to handle the units on the site and make them in the factory. But as far as the building itself is concerned, it is a great advantage to have such jobs done in a factory. An appreciable amount can be done;

electric wiring can be installed on partitions and plumbing can be fitted into sink units. Doors and their handles and other fittings can be put together, and the final finish applied, before they are incorporated in the building. False ceilings can be supplied, complete with fittings for lighting, heating and ventilation, and with connections for services in the building. Standardization of dimensions is of course vital if the maximum return is to be gained from such procedures.

Plastics bring their own peculiar versatility and lightness to the building. They are quite easily moulded in fairly large sizes and are light and strong—the builder likes them because they are easy to handle. They stand up to the weather for long periods, although they are gradually attacked by oxygen and become brittle. If it is rare for them to be load-bearing, rigid plastics foams and foam 'sandwiches' are employed as insulation in walls and ceilings. Insulating foams can be generated by squirting the ingredients directly into cavities between walls. The use of rigid polyurethane foam, for instance, has been rising rapidly. PVC and other plastics are made into gutters and drainpipes.

Manufacturers have been able to produce whole bathrooms in one plastics moulding that can be lifted from a lorry by crane on the building site and lowered into place, and then connected up to the electricity, water and drains. Even complete houses can be made of plastics of various kinds. Several specimens have been built and lived in for a number of years; but they seem to spark off little enthusiasm. In the main, people appear to like what they have always liked as far as living accommodation goes.

That is not to say, however, that this sort of stability is all-pervading. In areas where life is more mobile, there will almost certainly be a growing market for light, demountable buildings produced in the factory, speedily erected on site, and quickly dismantled when the time comes. More than a third of the homes being built in the United States are stated to be of the type that can be hitched to the back of a lorry and moved to another part of the country when required. Light, modular buildings can be made on the assembly line, being almost complete when they leave the works, down to windows, plumbing and wiring. Transported by lorry to the site, they are swiftly put up there, whether it is classrooms, laboratories, medical buildings or homes.

The Shape of the City—Problems and Plans

The city is a perplexing compromise between society's desire to gather in large groups and the individual's desire for the privacy of his home. The problems of how to build a pleasant, human, urban mass environment is one that is currently exercising the best planners and architects around the world. Many of the older cities grew up around a focal point—a palace, a church, a river crossing. Often they were surrounded by a city wall, and although many European towns, for example, dismantled these boundaries in the eighteenth century to allow for further growth, the pattern of their centres was already determined—they were patterns that were expressions of the needs of the people living there. They grew organically, in unplanned profusion, with houses, shops, offices and workshops all cheek by jowl. Traffic moved through the town centre at the measured pace of the pedestrian or the horse.

With the coming of the Industrial Revolution and its large factories and massive housing schemes, and its trains, trams and motor vehicles, the cities took on new shapes and were subjected to new pressures. The agricultural workers who flocked to the towns to work in the new factories needed homes, and they got homes, of a sort: rows of identical brick dwellings, constructed not as an organic part of the city, but merely as the most convenient places to house factory workers. They were usually built in the easiest, most space-saving form, in straight lines of back-to-back dwellings with no focal point to their pattern and few concessions to finer human needs. Though the inhabitants managed to create strong community ties within these unprepossessing surroundings, the basic layout was so bad that later ages saw them become gradually gloomier, and finally turn into slums or 'twilight zones'. Dirt, noise and disease became constant companions.

Meanwhile, the rush of new forms of transport brought to the city people who wanted only to get through it in their heady progress from one point outside to another. To them the confused, congested streets of the city centre were an annoying bottleneck. As the city became more congested, more well-to-do workaday citizens removed themselves to more gracious sites on the outskirts, away from the rude turmoil and smells. As demand pushed up the price of land, living in the centre became too expensive and the fashionable areas were taken over by the imposing offices of large companies and government departments, and the financial and legal firms to service them. The slum areas were left to themselves. These trends are still apparent. Most cities have been dying from the heart, as people move further and further out into the suburbs in search of the 'best' areas, and the city centres are surrendered to the massive vaults of office blocks, empty at night and at weekends, or to 'ghettos' of decaying slums in which the poor live. Thus, while the overall proportion of the population in urban areas is increasing in nations throughout the world, the number of people actually living in the city centre diminishes. The suburbs are little better, existing as they do in a sterile limbo between town and country in which people cannot quite align their loyalties to either: products of the rapid growth of cities caused by technology.

Generally, working people have to travel from the outer rings of dormitory areas into work and out again, placing a great strain on the transport systems, and themselves. The frustrations of the city—among them the stresses of travelling to work amid crowds of other commuters, and chronically high levels of noise and air pollution, usually from motor vehicles—must surely make people bad-tempered, unfriendly, even unhealthy. There is some evidence for it, even though studies of such relationships are at an elementary stage as yet. As cities expand, the problems multiply. For example, whereas in 1963 most commuters into New York City travelled less than 24 km (15 miles) to work, it is estimated that half of them will have to travel 40 km (25 miles) by 1980.

The last great challenge that cities in the industrialized nations overcame was that of disease. Better public health engineering and better and more numerous hospitals prevented recurrences of such dreadful epidemics as those of cholera and typhoid that came to Britain in the nineteenth century. The developing countries are still fighting on the same front. The next great challenge is that of finding efficient transport for journeys, both within the suburbs and outside, that will not stifle the city in the way that the car is doing. Present modes of public and private transport cannot cope (see page 177). The cities that were founded in large part before the mass-production of the motor vehicle have little hope of reaching even a good compromise between the needs of transport and the shape of the city, unless they ban vehicles from their centres and adopt some radical system that can be fitted into that existing shape, or raze the existing centre and rebuild to suit the motor car. New cities have a much better chance of adopting the right technology from the start.

The main problem for the planner is to keep dense human settlement within bounds while preserving the better elements of rural life within reasonable distance of the inhabited areas. Few people, it seems, fully enjoy living in the city all the time. The mass exodus of cars that winds its way out of a large city on any fine weekend is evidence that many people feel the need to escape from the soulless man-made landscape and get among the trees and fields every so often.

The closed 'cell-like' structure of earlier cities is not open enough for the highly mobile twentieth century. Present-day roads and railways are barriers that divide sections of the city, instead of linking buildings, as streets used to do. New cities—or rebuilt city areas—ought to recognize these factors. Instead of being built round a simple city centre they may need to be planned with dispersed functional centres along transport routes. The theme of many new towns—and indeed of many rebuilding schemes—which are trying to take greater account of people's needs in their living environment, is to combine the best of town and country in a different way from the current suburban sprawl. The British New Towns represented the first attempt to build on a greenfield site a new urban area combining all the necessary features of social and natural surroundings. The experiment continued, and twenty-two were built or started between 1947 and 1968, with government aid. The photograph (Plate 12.3) shows one of the latest, Cumbernauld, in Scotland.

A new city in future may be very different from the present, and be more a series of self-contained villages, subdivided into neighbourhoods of a few thousand people, each one its own focal point, divided but linked to a central core by untouched countryside. Where possible, large tracts would be forest and lakes. An informal grid of roads at, say, kilometre intervals, would cater for maximum use of the motor car. Within the grid where would be only local roads for slow traffic.

This type of city works on the principle of a centre for each level of community life. The neighbourhood has a small centre, oriented to local families and their younger children. Each cluster of neighbourhoods surrounds a village centre where more general services are available. In turn, the clusters revolve around the city centre which has a natural bias towards commerce but still retains some residential flavour.

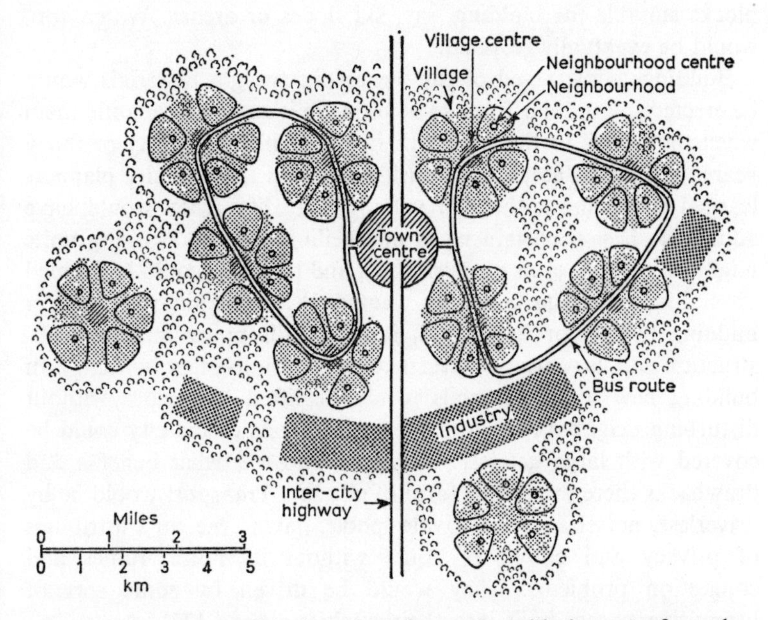

Diagram 12.2 One possible shape for a future city, with the centre for each level of community life—neighbourhood, village and town. The villages would be self-contained but linked to the town by untouched countryside

A radical American plan for an experimental city in Minnesota spells out some of the technical possibilities of a new city. It would ultimately house 250,000 people, and the industry and commerce needed to support them, in a densely populated area surrounded by an open space perhaps one hundred times as large. The city would try to incorporate radical ideas learned the hard way in other cities. It would be built with many of the services underground. In its substructure would be heating plants and cold stores, reservoirs for snow and rainwater from the streets, stores for building materials, power, telephone and other lines, and pipelines to carry out solid wastes. A network of communication cables would link central computers with terminals in various parts of the city and would permit inhabitants to communicate by telephone and other means. Vehicles would travel through roads in the substructure, their fumes being carried away by fume sewers to scrubbing and processing plants in the open area. Ultimately, all wastes would be recycled: solid wastes could be compressed and processed into blocks suitable for building, say, ski slopes or arenas. Water, too, would be eventually recycled.

Buildings constructed of the newest lightweight materials would be erected in modular form so as to make it easy to dismantle them when they came to the end of their useful life—in twenty or thirty years, or sooner, as people's wishes changed and the city planners learned more about the real value of city life. They would have adjustable floors, curtain walls and ceiling heights. Housing units might be precast, even pre-furnished, and the units could be juggled around like building blocks as wanted. When the time came for a building to die, or be altered, it would disappear into the substructure and its re-usable parts incorporated into other buildings. In building new ones, materials would be lifted into place without disturbing neighbouring ones. Certain sections of the city could be covered with large geodesic domes to find out what benefits and drawbacks there are to this form of city life. Transport would be by driverless, noiseless, semi-private 'pods', having the car's attributes of privacy and choice of route without its noise, fumes and congestion problems. They would be driven by some sort of propulsive system built into their track (see page 177).

Further in the future, the city may be unlike anything the twentieth century has yet disclosed. Visionaries foresee huge towers, each a city in its own right, holding many thousands of people, and rising perhaps 1·6 km (a mile) into the clouds. Others talk of a city on the ground covered with an immense dome of glass or plastics to maintain an equable climate within. In certain areas such a dome would eliminate snow removal problems and would make heating less costly, and it might pay for itself in ten years. Others envisage a sea city, built on stilts in the shallower waters around continental coasts. One design would be made up of pyramid-like neighbourhood units, each holding about 5000 people, linked to each other by causeways. Still others envisage a city like a huge ant hill, each dwelling a small box, stacked higgledy-piggledy on top of each other. These are attractive ideas where the money and the land are forthcoming. But in existing cities the opportunities are limited, and planners have to make the best compromise they can with the help of technology, given the existing city structure.

The city represents a wholly artificial environment, in which a growing proportion of the world's peoples have to live, and we are still extremely ignorant of the effects that such an environment has on its inhabitants. Massive urbanization is a comparatively recent phenomenon. By and large, mankind has only experienced living in large agglomerations like this for three or four generations. Already certain disturbing features of city life are becoming evident. Quite apart from the appalling conditions in which large proportions of their people are condemned to live, every city dweller is subjected to psychological tensions caused by coping with life among large masses of mainly unknown people. One can draw unpleasant parallels with the aggression shown when animals are kept in over-crowded conditions, although the validity of the comparison is not altogether certain. Certainly crime, and especially violent crime, has been rising sharply in cities, faster than in rural areas. So have vandalism and wanton violence. So have certain psychosomatic afflictions. Is it possible that these factors mark the emergence of a new breed of 'city man', in whom the better human qualities are suppressed by the effect of the city environment, and the worst brought to the fore? Does he have to vent his frustrations in these ways because of what the city does to him? Will these de-humanized features become more pronounced as selection in the artificial city takes its course? Only time will tell what the man-made environment is doing to Man.

13 · Technology and the Planet: Is Man Safe?

All Good Things Around Us: The Environmental Balance

Until now Man has plundered, pillaged and polluted his planet with careless abandon, thinking only of his short-term gain. He has used technology as a crude instrument with which to wrest food and materials from it, as though his were the sole claim on its resources. He has been too blind to see that he depends on the well-being of the environment, including all its other forms of plant and animal life. Some 350,000 different kinds of plants and more than a million kinds of animals—about 660,000 of them insects—are known to inhabit the earth, in company with him. A handful of soil contains hundreds of millions of bacteria, protozoa, fungi and other lowly forms of life on which depend the health of the environment and hence of Man himself. Among them are those that decompose dead plants and animals—including Man—returning their nitrogen to the soil; those that convert sulphates in sea water to hydrogen sulphide, which can then be used by other forms of life; and those that consume leaves, bits of bark, branches and so forth, producing carbon dioxide, water and heat. All of them, including Man, share a limited amount of resources found in the air, water, earth and sunlight. Ecology, the study of plants and animals in relation to their natural environment, is not a new science, but Man is fast discovering that because of his multifarious activities it must be applied to him just as much as to the lesser forms of life.

The basic unsettling factor is that technology imposes on the world artificial environments of Man's choosing, designed for his comfort and convenience, and by doing so materially upsets the complex balances of Nature. The worst aspect is to be seen in the grim devastation that often accompanies industrial progress: smoke-laden air and smoke-grimed buildings; filthy streams, rivers and lakes; abandoned, derelict factories and mine workings; spoil heaps of discarded waste. Many countries bear these industrial scars.

However, there are many other consequences of technology, frequently not so easily recognizable and yet potentially as harmful in the long term. Man has materially altered the whole form of large areas of the world by building cities, roads, reservoirs, airfields and other constructions. These have driven out much of the wild life of these areas, and altered the subtle interaction of sunlight, wind and rain with the earth; fundamental processes that have shaped the earth into its present-day form. Man has shot and butchered animals that he did not choose to rear and has deliberately destroyed them by chemical attack. The ubiquitous motor car also takes a heavy toll of animal life.

The rising use of artificial fertilizers, herbicides and pesticides has contaminated living things with chemical residues. Since the mid 1940s, when the large-scale dissemination of organo-chlorine pesticide compounds began, something of the order of a million tons have been scattered on the earth. Before 1940, pesticides were fairly simple compounds, such as lead arsenate, and some plant-derived substances such as pyrethrum. But large numbers of more complex compounds, by-products of military work in the Second World War, became available when the war ended. Today there are many such compounds, including DDT and other chlorinated hydrocarbons, organic phosphates, synthetic pyrethrum and nicotine. Some quickly break down in the soil into harmless fragments: others, like DDT, persist. Because of Man's profligate use of it as a 'cure-all' DDT is now one of the most widely-distributed synthetic chemicals on earth. It has been found in air, rainwater, soil and in animals all over the world, including Arctic penguins which live far from any treated areas. It occurs in food, in fish and animals, and in mothers' milk. Though it is probably the substance that has killed more insects and saved more people in its day than any other, serious doubts about its long-term safety have come to light in recent years. It is known to block the fundamental energy-releasing process in the living cell, the universal mechanism by which all forms of life draw energy from food through complex chemical cycles. On a primitive level, the accumulation of DDT in the oceans may already be affecting the phytoplankton (minute marine animals) in limited areas, and interfering with the photosynthesis they perform. Though it has not yet been proved to be a direct danger to humans, there are good grounds for believing that it has been responsible for large numbers of deaths of fish and lower animals, who accumulate it in their bodies from their prey. Realization of the dangers brought very strict control on the use of DDT and other persistent pesticides throughout most of the industrialized world from the late 1960s onwards. Other nations have to decide whether the possible risks outweigh the benefits of use, but unfortunately the results will not be confined to Man alone.

Man has converted vast areas of former agricultural land into dust-bowls through over-grazing and over-cropping, while he has poisoned other tracts by over-enthusiastic irrigation which has raised water tables, waterlogging the ground and making it saline. He has polluted the air by discharging combustion gases and dirt in vast quantities during the burning of hydrocarbon fuels. He has spread radio-active fallout across the globe, so that every child born since the Second World War bears some man-made radio-active strontium in his or her bones.

Man has polluted the seas and waterways by dumping wastes of all kinds, including radioactive rubbish, industrial wastes in many noxious forms, nerve gas and sewage, posing major threats to aquatic life. The actions of certain familiar substances that go into the sewage system in the normal course of events may be far from expected: for instance, certain detergents that are not completely destroyed in the sewage treatment frothed and foamed across the rivers and canals of the industrial world not very long ago. Then again, the level of oestrogen from birth control pills in rivers has risen considerably in certain areas, passing into the raw drinking water supply for other communities, with unpredictable effects. Apparently beneficial substances, such as fertilizers, can also have unpleasant side-effects. Nitrates from fertilizers may build up to dangerous levels in foodstuffs; they may also leach into ponds and lakes, encouraging the growth of unwanted algae which remove oxygen from the water, harming fish and other forms of life for whom oxygen is vital.

Man has crisscrossed the world with electromagnetic radiations from all kinds of sources, ranging from X-ray generators and microwave cookers to television transmitters, knowing little about their action on the body under these circumstances. He has unthinkingly subjected himself to extreme sensations of acceleration and deceleration, speed, noise, light and temperature, perhaps outside the competence of his body to withstand and beyond his competence to judge harmful or not. He continually consumes, either on purpose or by accident, great numbers of new chemicals as food additives, medicines or residues of agricultural chemicals in food.

All this, and more, has been done without knowing anything, and thinking little, of its long-term effects. Man's understanding of the environment is far too primitive to predict all the consequences of these actions. Some things—like DDT or air pollution—may make their presence felt rapidly enough. Others, like radioactivity, may take much longer. The others may or may not have an effect, and it may be beneficial, neutral or harmful in varying degrees. Whatever the consequences, Man can no longer escape. He is a prisoner of technology on a small planet of limited resources and limited capacity. He has to live with his wastes.

Man, as we have seen, is a comparative newcomer to the evolutionary scene. He arrived on earth something like two million years ago. Physically and mentally he takes millenia to adapt, through evolution (or 'natural selection') to changes in the environment. Living organisms in general evolve to suit their environment. If that changes, they must change, or perish.

Now Man is changing the environment on an unprecedented scale, so much so that large parts of it would be unrecognizable to someone living as comparatively recently as one hundred years ago. Man is still subject to selection, except that we can no longer call it 'natural', since Man is to a large extent creating the environment in which this selection takes place. There are several long-term dangers. One is that the constant strain imposed by the biological processes of evolution trying to keep up with the rapid man-made changes in the environment will lead to unpleasant, even lethal, psychological and physiological results. It is unlikely that Man can adapt with sufficient rapidity, and he is continually falling further behind as he changes his surroundings more radically.

Another danger is that by altering the genetic constitution of the human race Man will become more susceptible to natural forces. For example, increasing numbers of people physically incapable of normal life are being sustained by means of technology. They are liable to suffer considerable mental stress in trying to lead normal lives. At one time, they would have died. Now they can live, and have children. If their genetic defects are transmitted, a growing proportion of the population can become incapacitated in this way.

The path of evolution is strewn with the relics of species that were unsuccessful in the harsh strife of natural selection: species that initially evolved too quickly and too well, but in a way that increasingly restricted their choices of alternative ways of life until some change in the environment wiped them out. Some species, though, find a safe niche in the evolutionary edifice and can survive unchanged for aeons. Ferns have hardly altered in 350 million years. The lowly horseshoe crab, the opossum and the curious duck-billed platypus have existed much in their present forms for millions of years because they have found their niches. Their individual effects on their surroundings have been slight. Yet Man, in far less than two million years, and mostly in the last two hundred, has altered his surroundings, and the world's plant and animal life, on a tremendous scale. And he has done it—until now—in blissful ignorance. How can he be sure his niche is really safe amid all this upheaval?

Towards a Man-Made Climate?

The earth's climate—and hence the well-being of all forms of life—depends on two thin, fluid films around the planet: the atmosphere and the oceans. The atmosphere shields life forms from deadly radiations from the sun, while allowing them to bask in its warmth. It is a delicate balance. Too little sunlight and Man will freeze; too much and he will get skin cancers and shrivel. Besides providing Man with air, the atmosphere keeps the earth from losing its warmth into space through heat radiation. The oceans are vast stores of heat, energy and life.

By his pollution of the atmosphere and the oceans, Man is beginning to affect the natural cycles that occur in these two huge reservoirs of air and water, and upset the balance. One way is through his burning of fuels on a massive scale (see page 148). The liberation of large amounts of heat in this way will warm up certain areas of the earth directly, but there are other consequences. The combustion process also yields visible grit and soot, as well as steam and the invisible gases—carbon dioxide, carbon monoxide, sulphur dioxide and nitrogen oxides—and small amounts of many other substances. Over and above the immediate harm done to people, animals, and plants by this noxious mixture, there may be a long-term influence on climate which is only just beginning to be realized (Diagrams 13.1 and 13.2).

Carbon dioxide absorbs no appreciable amount of the radiation reaching the earth from the sun, but it does absorb longer-wave

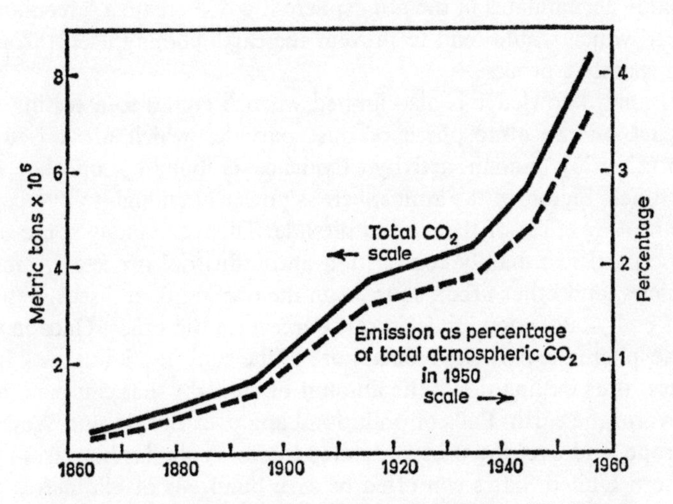

Diagram 13.1 Carbon dioxide from the combustion of fuels in the world, 1860–1959 (Royal Commission on Environmental Pollution)

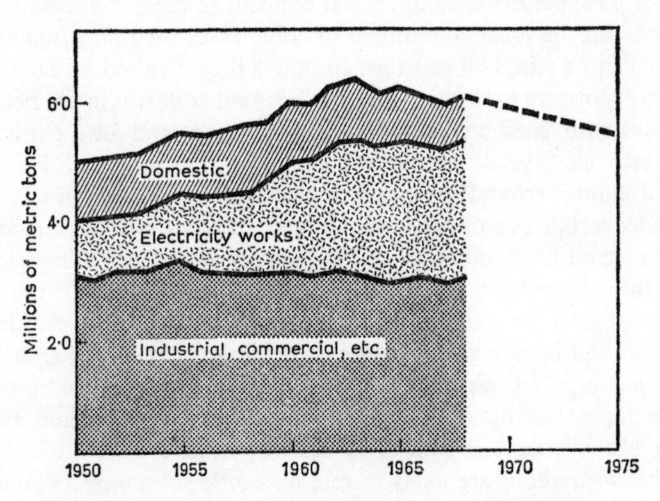

Diagram 13.2 Sulphur dioxide emissions in the United Kingdom, 1950–1968, with predictions to 1975 (Royal Commission on Environmental Pollution)

radiation by which the earth gives off heat into space. If carbon dioxide accumulates in the atmosphere it could create a 'greenhouse effect' which would tend to prevent the earth cooling itself through this radiative process.

Human knowledge is also limited when it comes to assessing the impact on the atmosphere of dust particles which are raised in profusion by human activity. Evidence is building up that the 'dustiness factor' of the atmosphere is powerful enough to overcome the heating effect of the carbon dioxide. There are many sources of these particles: mainly combustion and industrial processes, motor vehicles, and other urban sources on the one hand, and seas, deserts and badly-cultivated lands (rural sources) on the other. Once in the atmosphere the particles reflect more of the sun's radiation back into space, thus cutting down the amount of sunlight that can penetrate to warm the earth. Palls of pollution hang over Britain and Western Europe, and over the western United States. A similar haze from the eastern United States can often be seen hundreds of kilometres out over the Atlantic.

Besides forming a barrier to the sun's rays, the particles can act, too, as nuclei on which ice crystals, or fog or cloud droplets, may form from supersaturated air. What happens to these condensations depends on the local conditions. In some cases they may manifest themselves as rain, hail or snow. In others they may reduce rainfall, keeping moisture in the air as cloud. The lead additives in car petrol combine with small numbers of iodine molecules and form particles on which ice crystals can grow, so that there are extensive ice crystal plumes around a number of large American cities in winter.

Under certain conditions local fogs and mists can also be produced by the steam given off from cooling towers at power stations, and even from large heating schemes, especially in winter when the air cannot absorb the water. Although it cannot be said with certainty that this will in turn produce more low clouds, it seems likely.

Higher up, jet aircraft emit large quantities of combustion products; carbon dioxide, water vapour, nitrogen oxides, and dust. Their effect is far less serious in the troposphere where turbulent weather formations are likely to remove exotic substances relatively quickly than in the rarefied, more stable stratosphere. It may be that such substances disturb the balance of the stratosphere's layer of ozone that protects Man from harmful ultra-violet radiation from the sun. Accurate prediction of long-term effects is as yet impossible. Although the atmosphere's natural processes are powerful self-cleansing agents, they are being grossly overloaded by the wastes that Man consigns to the air.

Man's appreciation of his influence upon the oceans is even more scanty. His pollution of them by oil slicks, waste substances, and waste heat, has endangered much marine life. Oil slicks, in addition to fouling and killing, reflect less of the sun's rays than the water around them, and alter the forces of interaction between the moving air and water. If oil slicks become widespread, as appears likely, they could alter the interaction between the oceans and the atmosphere on a considerable scale. No one can say definitely what the effects would be.

It has been one of Man's cherished dreams that he would some day be able to control the weather. The method of attempting this feat has changed: once it was sacrifices, rain dances and libations. Now Man may be doing it unknowingly with his left hand, while trying to do it deliberately with his right. The long-term aims are to prevent or ameliorate natural disasters, so as to avert drought and prevent hurricanes, typhoons and storms scything their way across the land, killing and destroying. The early efforts of the Second World War were to disperse fog by setting up oil burners around airfields. The first major advance came in the mid 1940s when seeding of cloud and fog was demonstrated. So far these techniques have been confined to small areas, and it can be shown that seeding a small, isolated cumulus cloud with, say, silver iodide will cause light, and sometimes heavy, showers. It is relatively easy to clear larger areas of thin, cold cloud and fog. But the effect is small compared to the scale of the natural processes involved: perhaps a 10 per cent increase in precipitation over a small area. The search is on for techniques that will affect greater areas, and will suppress hail, lightning, and hurricanes.

For the most part, it is still a matter of research: collecting data, finding out how the weather works, and constructing theories for its behaviour. The problems are formidable. Meteorologists are trying to improve their understanding, and hence their weather forecasts, by building up mathematical models—sets of equations describing how the atmosphere works—whose behaviour can then be used to predict actual effects. Unfortunately these models have to be extremely complex: 200,000 or more variables may have to be written in to specify the structure of the global atmosphere at any given instant. Less ambitious models, however, can improve the accuracy of short-term forecasts, enabling meteorologists to make forecasts a week ahead with the accuracy of next-day forecasts. Computer calculation is vital. But calculation is invalid without reliable data, and at present there are pitifully few data-gathering stations over the majority of the area of the land masses, and still fewer over the two-thirds of the globe that is covered with water.

Efforts to discover how weather patterns form are therefore hampered even in the basic exercise of data collection. Yet only when the interactions of sunlight, air, earth and water are understood can Man hope to discover how his activities are altering them. In the meantime, attempts to make large-scale changes to climate deliberately are foolhardy, especially against the background of Man's accidental modification. Yet in 1959 a Russian proposal was put forward to dam the Bering Strait and use nuclear power to pump cold water from the Arctic Ocean through the dam into the Pacific. Cold water so removed from the Arctic would be replaced by warmer water from the Atlantic, perhaps enough to cause important climatic changes in the Arctic. In 1960 another Russian condemned the idea, because even this slight warming could, he said, decrease the precipitation over the Soviet Union enough to wreck the country's economy. At the moment, Man cannot predict the course of the natural processes of the atmosphere and climate, let alone the effects of his intervention.

Discarded Wealth

The middens of the planet's earlier inhabitants make happy hunting grounds for archaeologists. The rubbish heaps of the twentieth century, however, may prove burial mounds for society, unless the wastes can be used more constructively.

What do we discard? Primarily, burnt fuels, materials that have been incorporated into Man's constructions and machines, and human beings. The latter are little trouble, although many races insist on devoting an inordinate amount of space to storing, or at least laying out, the remains of each body. Under the action of animal, vegetable and chemical processes, Man's ingredients are broken down and returned to the environment: dust to dust. And each body contains valuable resources that future generations will need. A 70 kg (154 lb) man contains roughly 41,000 grammes of water, 12,600 grammes of fat, the same amount of protein, 300 grammes of carbohydrate, and 4·6 grammes of minerals.

However, the problems of processing Man's abundant waste products are much more acute. The average adult excretes about a litre of liquid, and an indeterminate quantity of solids each day. In the 'natural' state in a rural population these could be dumped in a garden, in the fields, or streams, for micro-organisms to convert into useful products. In the mass-urbanized, technological environment, that is no longer possible, although many cities still discharge their raw sewage into the sea with reckless disregard of the common good. Not only do human wastes have to be reckoned with. There are also domestic wastes, laundry rinsings, liquors from chemical plant, breweries, steelworks, and all the rest, plus variable and sometimes overwhelming amounts of rainwater. Wastes turn up from some unlikely places. The food processing industry, for one, contributes generously to the disposal problems. A chicken-processing plant, for instance, that prepares 60,000 chickens a day, produces in that day no less than 2·3 million litres (0·5 million gallons) of waste water from cleaning and processing, 3000 litres (660 gallons) of blood—which, together with the feathers, is sold as fertilizer—not to mention 3000 kg (6600 lb) of rejected birds that are sold to dogfood manufacturers.

Elaborate networks for collecting and plants for processing these wastes have had to be set up and maintained so that they can be returned in wholesome or useful form to the environment. In turn, an important industry has grown up supplying equipment able to cope with such repulsive mixtures, through combinations of chemical, physical and biological treatment. With the great increases in population, and in the scope and variety of technological activity, many disposal systems are becoming sorely overloaded. In most industrialized countries the onus of cleaning industrial effluents is being put more on to industry, so that wastes are at least amenable to treatment by the time they reach the public sewers.

At least liquid wastes can be disposed of relatively swiftly. Man creates far more intractable problems with some of his other materials. Some, like pottery and glass, are of ancient descent, and we delight in finding potsherds from prehistoric communities, or glass from Roman times. Some, like plastics, are modern and less delightful. As living standards rise, the mass of population will be able to buy more goods and discard more, more quickly. Thus in

1955, some 45 per cent of British cars were over ten years old, but by 1965 only 17 per cent were. In other societies turnover is more rapid still. The sheer volumes of waste are staggering. This year, people in Britain will throw away about 20 million tons of rubbish, and it will cost more than £60 million to store, collect and dispose of it. An average community in an industrial country produces about 100 tons a day for every 100,000 people.

As the open, solid-fuel domestic fire disappears, the nature of the rubbish is changing. It is getting lighter and more voluminous. Over the past twenty years its volume has doubled, though the weight has not altered much. Containing less dust and cinder, it is richer in paper, metals and plastics, and, notably, it brings an ever-increasing proportion of items too large for the dustbin—such as mattresses, washing machines, furniture, and pianos. In the United States, it is reckoned that the production of wastes of all kinds has nearly doubled since 1920. The estimated cost of disposal of the 360 million tons of solid wastes created annually costs about $4·5 thousand million.

It is an expensive business to collect rubbish and dispose of it: and still more expensive to attempt any kind of reclamation on it first. Much can be recovered if the local community is willing to bear the cost, but unfortunately the cheapest disposal method—still adopted in many parts of the world—is just to dump it. This is becoming increasingly difficult as the densely populated areas run out of dumping sites. Many communities have already filled the available holes and marshlands, or are coming near the end of their reserves of such sites. New York city, for instance, heaps 4·5 million tons of refuse into its marshes each year, and has made picnic groves, golf courses, and other sylvan delights from them; but it expects to have exhausted all the potential sites within nine years. Chicago estimates the same thing will happen there in three to eight years, and is planning, as one alternative, to pile the rubbish into a ski slope. Other cities face the same kind of problem, though they may not tackle it as constructively.

One possibility is just to go further afield, to lonelier areas. Another more constructive alternative is to use the waste to fill ugly scars left by strip mining of coal, minerals or clay. Other technical answers exist. Compacted pulverized rubbish, squashed into blocks and possibly consolidated by dipping them into asphalt or concrete, can be used as road ballast or building blocks. In France, such blocks have been shown to be as strong as concrete blocks but lighter. They can be sawn and nailed, and are good insulators of heat and noise. In time, the home might be equipped with a compacting press to squeeze its rubbish into this form. Coastal or riverside towns could use compacted blocks for such things as building up the shore along a river estuary. Much of London's refuse is pulverized at present, and much of that is taken by barge down to the Essex marshes and used as infilling.

Another course, of growing popularity, is to burn waste, though that adds to air pollution. London is adopting this method on a wider scale. The heat can be used for district heating, and to drive generators and produce electricity. Although more expensive initially, the cost of the extra plant can be offset by selling the electricity. In London's direct incineration plants, for instance, all rubbish is burned without prior sorting on mechanical grates. Ferrous metals are reclaimed from the clinker (itself a useful building material), and waste heat is used for electricity generation.

The fly ash left after pulverized rubbish or other finely divided fuels are burned, can be used for making cement and bricks, for road building, etc. It may contain iron, gold, aluminium and silver worth reclaiming given suitable processes. In other fields, iron ore dust from steel plants is fed back to make steel, while sulphur dioxide in waste combustion gases and sulphur from oil refineries can be converted to sulphuric acid by several processes, and then used again in its new form. At the moment, it is far easier to separate ferrous metals—by a magnet—than non-ferrous ones, which may have to be sorted out from the rubbish by hand.

Alternatively, shredded rubbish maybe from a grinder unit under the domestic sink, can be pumped to a combined sewage and reclamation plant. There are several difficulties in actually transporting the stuff. One idea is to mix it with water and pump it as a slurry. Vacuum tubes might transport the wastes from homes and other sources to the grinding works, but in all probability would be too costly except in high-density areas where they could be installed in tall blocks of flats, for example. Perhaps the most sensible course is to process the organic part of the waste biologically, producing compost for farming and horticulture.

Obviously, the ideal solution would be to use all waste material again, whether water, dust and grit, metals, glass, paper and rags, or plastics and other organic ingredients. Although such inventions as the self-destroying bottle or can which disintegrate once empty can make a small contribution to the problem, the bulk of waste materials cannot be conveniently removed in this fashion. Radioactive wastes are among the most difficult and dangerous of all (see page 147) and special methods of storage or dispersal have to be adopted.

Since our reserves of materials are limited, we shall be forced increasingly to re-cycle them, however great the cost. Today we are nowhere near that position, and it is generally cheaper to find and exploit new reserves than to adopt widespread re-cycling. The next twenty or thirty years are likely to see a great change in the situation.

Man His Own Enemy

Man is his own worst enemy. Even if he finds ways of mitigating the worst effects of technology, it cannot save him from himself. If he continues to reproduce so prodigiously, he will either starve himself to death, or will create such enormous pressures in the strain of trying to fit his brothers and sisters into the limited environment of earth that he will trigger off massive conflicts that will bring him eventually or quickly to the same fate. In nature, the numbers of a particular species are generally limited by the amount of space and food available. If too many offspring are born, they are destroyed, over the course of several generations perhaps, by lack of food, by natural predators, or by other forces of their environment.

Mankind is multiplying so fast that it is outstripping the available resources. Yet, because of its technological powers to alter the environment, and to dig deep into the earth for the resources of other ages, mankind has been able to stave off the reckoning, for a while. It cannot buy much time, for resources are becoming strained. Immediately, food is in short supply. Other shortages will make themselves felt. In the race between population and economic growth, the people are increasing too fast for many to feel any benefit at all. Soon there will be too little potable water and not enough houses, medical supplies, teachers, or land space. Even now, there are, on average, about fifteen people to every square kilometre of the earth's land surface: in one generation there could be more than thirty.

The population explosion began in Europe, where science and

technology were first applied to utilize the wealth of natural resources. This allowed the people in that area to obtain more food and materials, and thereby improve health and living standards. The same factors are now being applied across the world. Medicine has largely eliminated the chief killing and crippling diseases, such as malaria—mainly since the end of the Second World War—and as a result life expectancy in the industrialized countries has risen to between 67 and 72 years, as shown in Diagram 13.3.

In the developing countries life expectancy now lies between 30

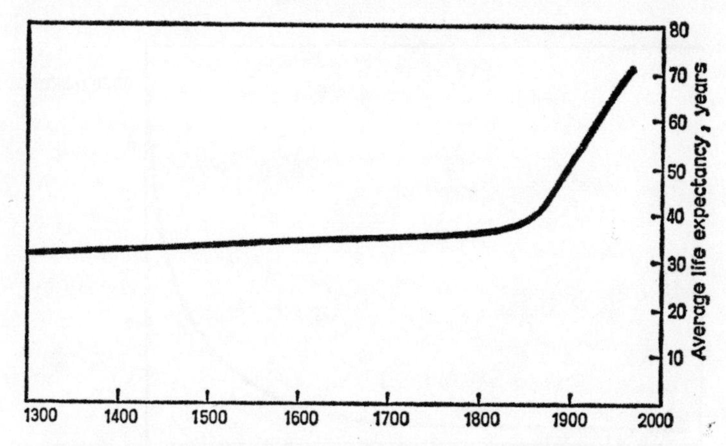

Diagram 13.3 Life expectancy has risen sharply during the past century, mainly because of advances in medicine (CBS Laboratories)

and 60 years, but the adoption of better public health and medical practices is enabling their peoples to live longer.

Thus, mortality in many Latin American countries fell in the 1950s at a rate five times that which England and Wales experienced when they were at a similar stage of economic development in the nineteenth century. If the birth rate has risen since the Second World War in western countries, and has remained high in the developing countries, it is more relevant to world population growth that medical practice has reduced the death rate, so that people are living longer. Mothers are being saved from death in childbirth to bear more children, and more children are surviving infancy and its formerly dreaded diseases to grow up and have offspring of their own. The average baby girl born in India in 1946, for example, could expect to live only 27 years. Now she will live to 48, through all the time she can bear children. So by fuelling the population explosion medical science and technology are exacerbating the problems.

There are believed to be about 3,600 million people on earth. The exact figure is unknown, because an accurate census cannot be taken. In any case the total rises every minute. Doubling every thirty years, the world population is due to reach 7,000 million by the year 2000. By the year 2040, which is within the lifespan of today's babies, it could be 15,000 million. Diagram 13.4 shows this exponential increase.

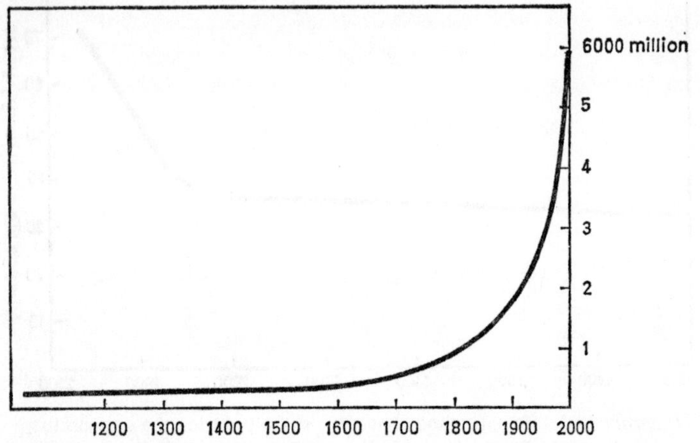

Diagram 13.4 The world population, increasing by leaps and bounds (CBS Laboratories)

Most of the increase will occur in the developing countries, which are least able to absorb the extra people. Already many are in desperate straits, and further large increases will strain their overtaxed resources severely. The population of Latin America is doubling now in 27 years; Turkey's in 23 (it will treble in 37); Central America's in only 20 years. Mankind breeding freely will bring catastrophe. As it is unthinkable that society would ever condone the principle of ruthless elimination that the Spartans are said to have practised—leaving weak and sickly babies to die on the mountain slopes—we have to adopt the alternatives of resorting to abortion, legalized or not, and birth control, in order to control population growth. The attitude to abortion is so emotional, and depends so much on a society's moral values, that it will be some time before it can be regarded as a freely chosen alternative. At least improved medical practices have made it a much safer operation for the mother than it used to be.

Birth control, however, is regarded with more favour, although it is still steadfastly opposed by certain religious denominations. At least the means are available. Medical technology has provided various methods, for both women and men, and will no doubt produce more now that the urgency of the problem is realized. Two important means, for women, are capable of mass production and are so cheap, highly effective, and apparently justifiable as regards risk. One is the plastic IUD (intra-uterine device) which, once inserted in a woman who can tolerate it, remains to offer protection against pregnancy indefinitely. This lowers the pregnancy rate among users, including those who have unknowingly ejected the device, to two to three per year for every hundred women. This is less than half the rate with conventional contraceptives under the best conditions. In India, where the control programme has got under way after a shaky start, more than 2 million IUDs had been inserted by the end of 1967, and nearly 1 million in Pakistan. But the drawback is that 20 to 30 per cent of women cannot tolerate the device or retain it. For them, and for many other women—in developing countries and industrial countries alike—a much more attractive proposition is the Pill, or synthetic steroids, taken orally. Now that prices are falling, bulk quantities can be obtained cheaply enough for government programmes. In both India and Pakistan, the Pill is starting to be used alongside male sterilization, itself a more drastic means than the well-tried condom for men. Current research into the process of fertilization may lead to more effective methods of birth control, without some of the drawbacks of present methods.

It is one thing to provide the means of preventing birth, and quite another to persuade people to use them. People normally want children. Besides, past generations grew up knowing that many of their children would die young. Since conditions are better, more children survive and grow up, in town and country alike. An Indian peasant and his wife, say, who must be able to count on grown-up children to support them in their old age, and provide for the rest of the family, can confidently expect that their babies, or most of them, will become adults, instead of succumbing to childhood disease or malnutrition. But it takes generations for these new truths to sink in and for people to realize that they do not have to have large families to ensure that a few survive.

But still, many people do now seem to want to limit their progeny. Preliminary sample surveys in some twenty developing countries have shown that substantial majorities of married couples want to restrain their childbearing. It is partly a result of direct or casual education: through films shown at permanent cinemas or by travelling shows; through radio programmes picked up on transistor radios; and through schools and health departments. Even those in remote villages realize that children no longer all die in infancy, that sweeping epidemics are no longer inevitable, and that a medium-sized family will provide enough adults to care for the grandparents in their old age. They appreciate, too, that the more children they have the more difficult it is to find them enough food.

Ten years ago, India was the only country to have adopted a national policy to promote family planning. By now some forty countries, containing about 80 per cent of the population of the developing world, are at work on one form or another of family-planning programme. Results are beginning to become apparent: India's births have been reduced in numbers by about 10 per cent, and in South Korea and Taiwan birth rates are falling. Perhaps the best example is Japan, where for over ten years the population has been held, until recently, at around about the 98 million level thanks to a vigorous government-sponsored programme that kept the birth rate among the lowest in the world.

While Man may do his best both to increase food production and restrain population increase—and the evidence seems to show that neither of these activities is yet being given the priority it deserves—he cannot do anything rapidly enough to prevent the next twenty years or so bringing him to crisis point. Regardless of the success or otherwise of birth control, the world will have to find food for twice the present population by the end of the century, and much more of the demand will arise in the developing countries where the food production in relation to population is already lagging (see page 228). Energy and materials will have to be found to supply the wants of all the extra people who will not have the money to pay for them; they will have to be housed and clothed, as well as provided with food and water. Work will have to be found for the unemployed. Medical services and teaching will have to be provided. There will be public health problems of unprecedented scale. By the start of the twenty-first century three out of every four people in the world will probably be living in a city and, very likely, it will be a city swollen by ramshackle 'shanty towns' of rural immigrants. This would spread the man-made environment and its attendant evils still further. At present, three-quarters of the population of Calcutta live in shacks without tap water or proper sewage control: during the monsoon when the waters rise they are obliged to wade through their floating excrement to get around. But even disregarding the horrible conditions in which many will have to live, the activities involved in meeting even their basic needs will impinge yet more grossly on the environment. And when it is a matter of choosing between the immediate well-being of Man and the well-being of the environment, is there any doubt which Man will choose?

14 · Technology: Healer and Destroyer

Engineering with Man

Although Man is never likely to achieve immortality, there are many circumstances in which technology can do battle with death itself, since machines can sustain life in people who would otherwise die. Death comes only when the machine is switched off. Other machines and instruments do more than prolong life, as they can restore a dying person to full health by pumping and oxygenating blood outside the body while surgeons repair a damaged heart. They can filter poisons from the blood of people whose kidneys are failing, and not only save their lives but enable them to go on earning a livelihood. Machines and sophisticated instruments can breathe for people with paralysed chest muscles; they can give the deaf a degree of hearing and enable the blind to perceive obstacles in their path; they can also enable a legless person to walk, and relieve the distress of incontinence.

Science and technology have achieved this almost entirely over the past hundred years. Before that the doctor had little to help him but his stethoscope, his scalpel, and his experience. There were no antiseptics to kill germs, no anaesthetics to eliminate the pain of surgery—and indeed, little understanding on the part of the medical profession of the causes of diseases, how they were transferred, or how they could be cured. It was a hit and miss affair. Patients were treated with a mixture of faith, hope and superstition, tinged with apothecaries' magic, traditional remedies using natural extracts such as quinine for malaria and more bizarre 'cures' involving ground-up insects and vermin. Such treatment was often more hazardous than letting the disease run its course, and a patient was more likely to die if he went to a hospital than if he stayed at home, such were the dangers of cross-infection.

Now the doctor has at his command some of the most sophisticated technical apparatus in the world, and the products of research-intensive industries such as pharmaceuticals and electronics, with which to fight against Man's diseases and disabilities. In the last two decades, the most notable factors have been the rise of the technology of bio-medical engineering, and of the pharmaceuticals industry. With the aid of bio-medical engineering Man can now to a great extent repair damage done to himself, and with the aid of drugs can destroy within himself the hostile micro-organisms that cause disease. Bio-medical technology, reinforced by materials technology, and other disciplines, has provided devices and machines that can assume the roles of parts that are deficient or absent in the human body. The diagram shows the more important ones.

Man-made joints, arteries, tissues, limbs and other synthetic replacements can make life more bearable for those who have been deprived of perfect anatomy by genetic chance, disease, or accident.

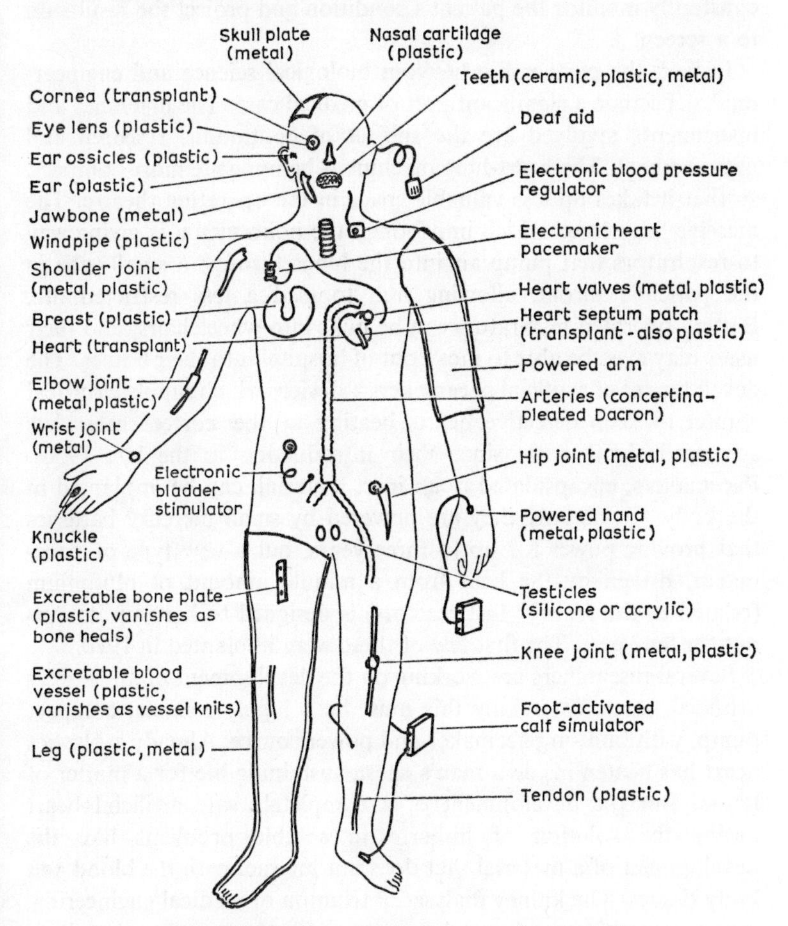

Diagram 14.1 Man-made replacements for the body

Although wood, metal, bone and other materials have been used to patch up the human frame since before the birth of Western civilization, new surgical skills, backed up by greater technical knowledge and the development of more suitable materials, have brought prostheses to many more people who need them in the last thirty years or so. Plastics have proved particularly versatile in many roles: as arteries, implants for joint surgery, as corneas for eyes, and as replacement heart valves. Operations involved in inserting synthetic parts into the human frame—particularly open-heart surgery—may take many hours, during which time machines and instruments constantly monitor the patient's condition and project the results on to a screen.

In fact, the partnership between biological science and engineering has become a significant part of medical care. The machines and instruments involved are the subject of continuous research and improvement. The heart-lung machine is being made more compact, so that it takes up less valuable space in the operating theatre. The massive 'iron lung' which imprisoned the polio victim is giving way to respirators that pump air into the lungs through a small tube in the patient's throat, allowing him to lead a less restricted life. Battery-operated respirators can be built into wheelchairs, and their users may even be able to move out of hospital into their homes. The development of artificial pacemakers—devices which supply electrical stimuli to keep defective hearts beating at the correct rate—has advanced significantly since their introduction in the late 1950s. Pacemakers, encapsulated in an inert material, can be implanted in the body. Commonly they are powered by small mercury batteries that provide power for up to three years, but a new type of pacemaker, driven by the heat from a minute amount of plutonium (related to the fuel for fast reactors) is designed to keep up its output for ten years. The first few of these were implanted in 1970.

Several researchers are working on the development of an entirely artificial heart. Essentially this must be a highly reliable, compact pump, with built-in pacemaker and power source. Already a plastics heart has beaten inside a man's chest, sustaining life for a matter of hours. But the development of a completely safe artificial heart awaits the solution of hitherto intractable problems like the development of a material that does not interact with the blood and body tissues. The kidney dialyser, a triumph of medical engineering, is an expensive and cumbersome apparatus, and research is constantly seeking ways of making it cheaper and more compact. Some have been devised which can be operated by the patient himself, and are compact enough to be fitted into his home bedroom. The ultimate aim is to provide a portable unit which could be worn on the arm, perhaps, or a waist belt, or even implanted. Research is being conducted on the possibility of artificially dialysing the blood in the case of renal failure within the body; the patient would simply swallow a small capsule containing the necessary chemicals.

Matching advances in the technology of surgery have been made possible by improved materials, for instruments, sutures, implants, and so on; better packaging (such as sterile plastics packs of swabs or disposable instruments); and improved sterilants and anaesthetics. More highly skilled anaesthetists also enable the surgeon to achieve a higher proportion of successful operations in serious cases. Among the devices the surgeon can now employ are cryogenic probes cooled by liquid nitrogen to freeze out diseased tissue in the brain and elsewhere; lasers for cutting out tumours and for 'welding' back detached retinas in the eye; high-voltage X-rays and radio-isotopes for cancer therapy; intense ultrasonic beams to break up gallstones and kidney stones; and high-pressure oxygen chambers in which patients often have a better chance of recovery from such conditions as carbon monoxide poisoning and gas gangrene.

Experience, refined technology and better means of detecting what is going on inside the body have given surgeons enough confidence to probe deep into the brain—to destroy tumours, or remove haemorrhages, for instance—and to transplant eye corneas, kidneys, hearts and other organs, and now even nerves, from one body to another. Blood transfusions are almost without hazard as the blood can be carefully matched before the transfusion takes place. The matching process cannot normally be carried out with transplants. It is true that transplant experiments have been backed by the development of low-temperature techniques for storing transplant organs, and the discovery of drugs to suppress the rejection mechanism by which the healthy body tries to destroy foreign tissue—the same mechanism that destroys germs, but which works against transplants. The first heart transplant operation, performed in South Africa by Dr C. Barnard in 1968, was followed by a series of similar attempts, but they seemed to be unable to sustain the patient's life for much more than eighteen months.

Either, the body's rejection mechanism thwarted acceptance of the 'foreign' material, or the massive doses of immuno-suppressive drugs needed to control this mechanism left the body unable to combat infection. Even so, such is the rate of medical progress that the outstanding problems of heart transplants and artificial hearts will no doubt be solved within a short time, and Man will have succeeded in replacing or synthesizing one of his most vital organs, a feat unthinkable only a generation ago, and one that would smack of divine intervention to the medieval mind.

Man is also becoming wonderfully skilled at repairing his limbs and making technological substitutes for his senses. Electronics, and especially miniaturized electronics, has played a major part, with valuable roles for the technologies of pneumatics, hydraulics, and electrical engineering. Competence in all these fields, in materials and in others has led to the design of such devices as the hearing aid, and of various units that allow the blind to 'see' through a camera or other detector whose output is converted into audible or tactile signals. The design of artificial limbs has also benefited. The closest approach to natural limb movement comes when the artificial limbs or parts are triggered by electrical impulses from the brain to muscles in the stump which would activate the natural limb. Sensors for these tiny currents replace mechanical or other actuators which generally have to be operated by the patient making some other movement like shrugging his shoulders, or moving the other arm. For example, an 'electronic arm' for people who have had an arm amputated is attached to the stump of the upper arm and activated by sensors (electrodes) stuck on the skin over the main muscles, controlling arm movements, in the chest, back, and shoulder. Four small motors, driven by batteries that can be carried in a belt, power the arm to enable the amputee to carry out smoothly eight kinds of movement, including writing. The reverse process—providing powered electrical impulses to trigger bodily activity—is also valuable. It can be applied to make defective muscles contract, for instance, and to restore paralysed fingers with electrical signals applied through implanted electrodes. Although these devices have reached a high degree of technical perfection, they still fall a long way behind the real thing. Intensive work is being done on an international scale to design and construct artificial aids that are lighter, more precise, refined, responsive and more natural in appearance for Man's physical and psychological comfort.

Engineering of another kind, not involving machines, is also possible, but it seems to hold the promise as much of disaster as of comfort. It is 'genetic engineering'. Taken overall, it means that Man may soon have the powers to change his physiological make-up in ways which would either take thousands of years to accomplish by natural selection, or would not occur at all in the natural world. Thus it is now possible to alter the genes—the microscopic units, strung out along the chromosomes within the human cell that determine the control and transmission of hereditary characteristics from parents to children. At the moment the methods are crude: chemical attack or X-rays produce changes in this genetic material, most of them undesirable and many lethal. This is the basic reason for the fear of radioactivity, whether from nuclear power or nuclear weapons. When Man tries such methods out on plants, all the unwanted mutants can be discarded; but obviously that approach is impossible with people. However, there seems no reason why these methods—with greater knowledge and experience—should not be refined so that it becomes possible to alter specific parts of the chromosomes, and thus produce animals or people with enhanced, or suppressed characteristics. Theoretically, within perhaps twenty years it will be feasible to engineer super athletes, or intellectuals, or a breed of very strong but unintelligent labourers, or in fact any kind of person required by the genetic engineers or their masters.

Then there is the possibility of rearing 'flask babies' by fertilizing in a test-tube human ova with spermatozoa from carefully matched donors, and either implanting the fertilized ovum in the womb of a chosen mother (as is now done with animals for breeding purposes) or, further ahead, rearing the ovum in a 'flask', with the aid of an artificial placenta. The baby would be reared until it had grown to the requisite stage, and then would be 'born' artificially—that is, become an air-breathing organism, existing on its own blood supply, instead of that supplied through the placenta. Not only could manipulation of this kind make its mark on the creation of life through the traditional method. Scientists are also discussing the chances of eventually breeding new human beings from a few cells from the skin, or other parts of the body, apart from the germ cells (ovum and sperm) from which life is created at present. Such a technique, called cloning, which has already been demonstrated in plants, produces identical copies of the donor. Fantastic though it seems at present, cloned human copies could be made in the near future, in the opinion of some authorities.

Such are the bewildering prospects for the future if Man takes seriously his powers of tinkering with life. Yet there are problems of just as much magnitude that demand his attention immediately. These concern death. The one thing certain in any man's life is death. But death can come at any time, and science and technology have continually thrust it further back into Man's old age. Our forbears in the Middle Ages lived for thirty years on average. In the last hundred years, and more noticeably in the last fifty, life span in the western world has lengthened significantly and recently passed the Biblical span of threescore years and ten. Birth itself has always been a dangerous event, for both mother and child. Yet, thanks to improved gynaecology and obstetric methods more widely applied, improved nutrition and better health among mothers, the infant mortality rate associated with birth has been substantially reduced. In Britain, for instance, it has fallen over the last fifty years from 153 to 22 for every thousand births. Only thirty years ago, infections were the leading killers of childhood and middle age. Childhood scourges such as whooping cough, measles, diphtheria and tuberculosis left few children unscathed, even in the advanced countries, forty years ago. Less than a century ago, one person in four died before reaching the age of twenty. Now these and other diseases have largely been conquered, through better sanitation, cheap drugs, etc. The recovery rate for sick people has been improved through better surgery, radiotherapy, physiotherapy and chemical therapy with drugs. Today, only one out of thirty-three young men and women in Britain will die before reaching the age of twenty. The present killers, which were formerly obscured or prevented by these diseases and factors, strike later in life: heart disease, cancer and strokes. Accidents and violence loom large as the next major cause.

Yet so far has medical technology progressed that people need not necessarily die when their vital organs fail. Life of a sort can be sustained for long periods by respirators that keep the heart and breathing going, even though the patient is deeply unconscious, without apparent hope of recovery. In this case when and what is death? Is it when the heart stops beating? The heart can be started again with massage or electrical stimuli, and kept beating with the aid of a machine. Brain activity continues for some time after the heart stops. Is death the point at which this activity ceases? The old definitions are no longer precise enough. And if the body can only be preserved in its semi-living, unconscious state by machines, when should they be turned off, if at all, and devoted to some other patient who has a better hope of recovery? After a few hours, or one day, or a year? How much money ought to be spent on preserving the life of one person in such a condition, if there seems to be little hope of their recovery?

Should medicine concentrate less on saving a few lives at any cost, and devote itself more to curing non-fatal ailments that make life miserable and affect large numbers, such as arthritis, or even the common cold? Intensive biomedical technology, like any other intensive technology, is expensive. The cost of looking after one patient in a heavily-instrumented intensive-care unit in a hospital may be more than £400 a week—about five or six times that for the normal hospital patient. Even in the best units not all the patients recover. Who, then, ought to benefit? There are not enough places for all. Would it be better to spend less on such units, and more on mass health care for the population at large? Society has to decide, and the choice is a cruel one, whatever the outcome.

Searching for Signs within the Body

A great deal of medicine is involved in trying to discover what is happening inside the body. Simply, this may be done by observing the exterior signs: a rash, a swelling, dilated eye pupils, a high temperature. However, many things are hidden from the doctor's eye or hand, and medicine has to turn to technology to reveal the workings of the human frame. Before science was able to elucidate the nature of the reactions going on in the body in its normal and diseased state, the knowledge on which the doctor could base his treatment was scant indeed.

The well-known stethoscope enables the doctor to 'listen in' to heartbeat, respiration and so on more easily than with his unaided ear, but it is no more than a very simple acoustic amplifier. However, ultrasonic signals (see page 106), can also detect structures inside the body. They can disclose brain tumours or haemorrhages, and are particularly useful in examining accident victims with head injuries, and patients with heart conditions or with foreign bodies in the eye. Because they do not carry the same kind of risks as X-rays when applied to pregnant women, they are used in maternity cases to disclose the size of the baby and its exact position in the mother's body, giving the doctor evidence with which to plan the best method of delivery. The ultrasonic generator scans the mother's abdomen and an electronic sound-to-light converter changes the emerging ultrasonic signals into echoes on a cathode-ray tube screen, which is then photographed. The ultrasonic 'camera' can also measure the heartbeat of the foetus. Another ultrasonic instrument measures blood velocity in the body through the skin, without having to actually bare the blood vessel and insert a probe into it.

X-rays have been applied in the medical field almost since Roentgen's discovery in 1896, although their use has been wisely restricted after discovery of the damage they can cause. Their power of penetrating flesh and portraying bone structure beneath was one of their early marvels, and is a power that modern medicine finds just as valuable for such purposes as picturing fractures in bones, and discerning tumours and other growths, and various other internal features. In the recent past, X-ray pictures have been improved by intensifying the brightness of images with electronic image converters, and by linking the X-ray machine with closed-circuit television to display the X-ray picture on television screens—possibly in other rooms within one hospital, or perhaps in another hospital many kilometres away. The pictures can be video-recorded, to be played back later for a specialist, or a group of doctors, to view.

Another technique is to track radioactively 'labelled' compounds (see page 138), as they travel around the body, with the aid of various types of scanning device. Different isotopes can be injected for different purposes. Iodine-131 concentrates in the thyroid gland—more rapidly in a healthy, active one—and therefore can be used to discover whether the gland is functioning properly. Another, calcium 47, concentrates in bones, and can pinpoint many bone conditions. Scanning devices, starting with the Geiger counter, have been developed since the early 1950s to map the pattern of distribution of the isotope throughout the entire body, or in parts of it.

Ordinary light is also useful. Ingenious endoscopes using fibre optics (see page 54), allow the doctor to scan tissues inside the body. These ingenious 'light pipes' can look forward and sideways through a miniature scanning system powered by tiny electric motors. With these and other optical viewers, doctors can inspect the bronchae, for example; or, by passing a light pipe up along the femoral artery, they can inspect the aorta.

Heat can also provide information about the condition of the body. Thermography, heat mapping of the body with an infra-red scanning device (see page 57), allows the doctor to detect cancers of the breast and other areas, and to investigate disorders of the circulation, by picking out abnormally hot or cold spots on the body. An infra-red-sensitive detector—often an electronic thermistor, which converts heat into electrical energy—picks up the heat radiated by the body, discerning temperature differences as small as 0·2°C. A variation of this technique is to coat the skin with liquid crystals that change colour according to the skin temperature.

A very fruitful source of information lies in the body's electrical activity. The body teems with minute electrical currents, pulsing through brain, nerves, heart, muscles, and the inventions of electronics technology have helped doctors to make use of these natural signals for diagnosis. The signals can be picked up through electrical contacts (electrodes) carefully attached to the skin, and sometimes inserted in the body. The signals are electronically amplified and displayed on moving-pen recorders or oscilloscopes (Plate 14.1).

The electro-cardiograph detects the surges of current that accompany the heart's muscular contraction. The electro-encephalograph detects 'brain waves', the tiny currents that emanate from the brain during sleeping and waking. These signals, much less intense than those associated with the heart, could not be detected at all until after the invention of the thermionic valve, which can amplify electrical impulses millions of times. Indeed, not until 1929 did detection become possible, compared with 1878 for 'heart waves'. Disease or infirmity distort the characteristic traces drawn by these instruments, though doctors had to collect and analyse many thousands of traces before they could tell by experience what was normal, and learn to associate a particular trace with a specific condition. ECG devices are necessary for monitoring heartbeats in such extreme situations as during surgery, or in the hospital's intensive care unit. Once the heart signals are in electrical form they can be recorded or transmitted to any desired point. Thus astronauts are monitored even when walking on the moon, thanks to reliable miniature sensors and elaborate telemetry systems that send the information back to earth. Other transducers (see page 112) can detect and signal information about respiration, blood flow, temperature, muscle condition and other quantities, and the signals can either be monitored or retained in recorded form for later study.

The use of monitoring systems is growing. At first they were employed to keep watch on cardiac patients in hospital, but they are now being utilized for other patients, among them foetuses, new-born babies, and patients recuperating after surgery. Systems of the kind can be put under computer control, and can be arranged to provide the doctor with continuous information on the patient's current condition. His diagnosis and treatment are based on this, and data from the laboratory, where blood, urine, and other samples are chemically analysed. The laboratory is also becoming increasingly automated (Plate 14.2). Analysis units are able to take samples at rates of 300 an hour, process them, and print out the coded results or punch them on paper tape for inspection or for feeding into a computer. The results can be handed to the doctor in the ward within half an hour of the samples being taken, after being checked in the laboratory to ensure that there are no gross errors due to faulty equipment. The information can be passed between the operating theatre, intensive-care unit, laboratory and the head physician's office, as shown in the diagram.

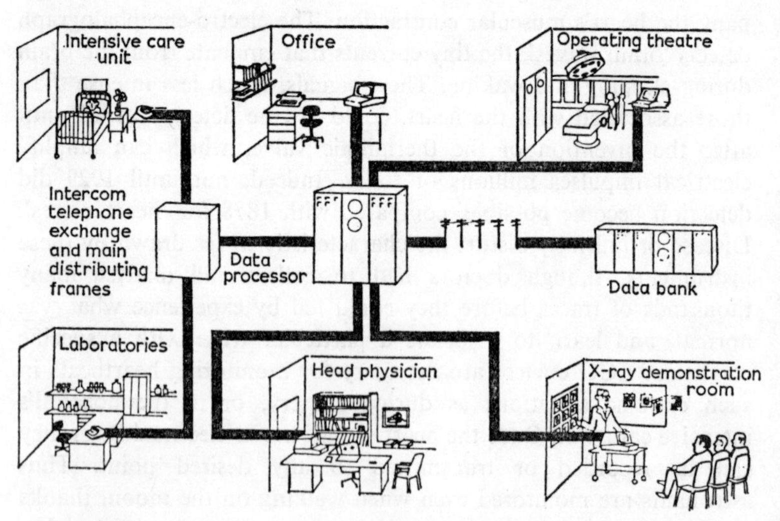

Diagram 14.2 A computer-based information system in a hospital

Once in the computer, however, the information can be transmitted to other computers, or terminals, wherever they are, and an important exchange of medical information takes place on an international scale, helped by computer-based information systems (see page 158).

The purpose of investigating a person's basic condition in these ways—through blood tests, X-rays and so on—is to find out if anything is wrong with him, and if so, what. At present, the machines do the tests, provide the data, and the doctor makes the diagnosis. But it is only a matter of time before computers are given the main responsibility for making the diagnosis as well. The ultimate aim is an entirely automatic system which makes a diagnosis once it has been fed with all the relevant information on a patient, derived from his case history, blood and urine tests, etc. The next logical step is to use such measurements as screening procedures to detect the early stages of disease among large sections of the population, like mass X-ray screening for tuberculosis.

All these detection techniques have given the modern doctor a far clearer insight into the actual nature of disease or injury in his patients, and have paved the way for automatic monitoring and, before long, automatic diagnosis. They can lighten the burden of the chronically overworked medical services that are struggling both to live up to rising public expectations, and to care for an increasingly ageing population (as better medical treatment enables people to live longer). Though the machine can never emulate the human sympathy and understanding that are so valuable to the patient, it can free medical staff from much routine work.

Chemical Defences Against Disease

Huge populations of micro-organisms occur throughout Man's environment in water, air and soil as well as in plants, animals and Man himself. Some are helpful and are used in fermentation in industry, for instance, to make alcohol from starch or sugar. Most are harmless, but a few can cause him grievous harm. Thus bacteria cause pneumonia, diphtheria, typhoid, tuberculosis, cholera and syphilis; fungi bring such diseases as ringworm and thrush; protozoa cause malaria; and viruses, so small that they can only be seen in the electron microscope, cause the common cold, flu, measles, smallpox, polio. If Man has natural defences against them, these may be overwhelmed from time to time. Until the end of the last century there was a great risk of dying from some microbial disease before the age of forty.

Since the founding work of Louis Pasteur and others, however, towards the end of the nineteenth century, a great many other defences have been brought into action as a result of scientific discoveries in microbiology, and latterly technology. Vaccination (inoculation), which was introduced to Britain in the early eighteenth century for smallpox, protected people against this and other diseases by stimulating the natural defence mechanisms of the body through injecting weakened doses of the causative micro-organism. From 1865 antiseptic techniques pioneered by Joseph Lister, using carbolic acid solution, allowed surgeons to open up the body and perform extensive operations without the wounds becoming infected with bacteria from the air and from the hands and instruments of the surgeon.

The next step was actually to attack and kill the micro-organisms while they were in the body. It had been proved that this could be done by certain natural products, such as quinine, obtained from the bark of the cinchona tree. But the first real test of the principle with man-made compounds came with the scientific discovery and success of Salvarsan, an arsenic compound discovered by P. Ehrlich in 1908 and used in a specific treatment for syphilis. Even so, until the mid-1930s the pharmaceutical industry was small, and it grew slowly. Most of its products had been used by doctors for generations, some dating back to the Middle Ages, and a few to the ancient Greeks. Most of those that were effective did not strike at the roots of the disease. They only relieved symptoms.

The change really came in the mid 1930s with the discovery of Prontosil, the first of the sulphonamides, and the first of the 'wonder drugs'. By showing conclusively that chemical substances could destroy disease-causing organisms without harming the body it set in motion a massive research and development effort. From this came the discovery and mass-production of the whole range of modern drugs from penicillin and tetracyclines for bodily illnesses to stimulants and tranquillizers for mental afflictions. In a short period, the drug industry has changed profoundly. The processing of roots, leaves and barks has been largely superseded by quantity production of highly complex chemical and biological products to the most exacting standards of purity, stability, quality. (Plate 14.3 shows a typical pharmaceuticals plant.) The production facilities are backed by an equally effective worldwide marketing organization. Large drug companies, like other manufacturing companies, are international both in outlook and operations. World pharmaceutical sales have risen by an average of 10 per cent a year since 1950 and in the future are expected to go on rising almost as fast. The diagram shows the position of the British-based industry.

Companies have to operate on a large scale in order to support the massive research and development programme that goes into establishing a new drug. For every successful compound there are on average 5000 others discovered or created that finally have to be discarded because they are unsuitable in some way. Once a promising

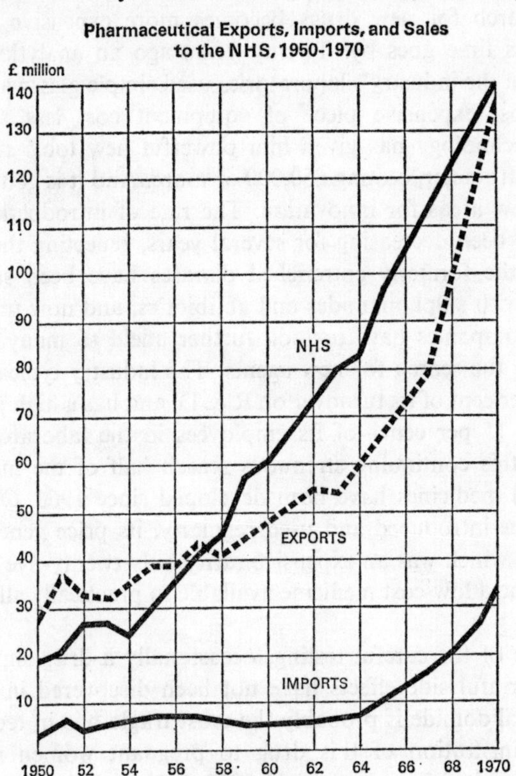

Diagram 14.3 British pharmaceuticals exports, imports and sales to the National Health Service (The Association of the British Pharmaceutical Industry)

compound is found, it is investigated further to determine the probable type of action it will have, and the easiest and most effective form it should take. A wide variety of biological tests are conducted to see how it works and if there are any side-effects, and to establish the safe dose. These include tests on healthy human volunteers, often members of the research team, to enable the researchers to check and confirm their findings, and observe the volunteers for a number of years in case of any long-term side effects. When they are satisfied that the drug works, clinical trials have to be carried out. These will last for five to ten years and involve as many as 10,000 people. That explains why the complete cost of a new medicine that reaches the public, after five to eight years' work, is frequently more than £2 million.

The search for new drugs becomes more expensive and more difficult as time goes by. Twenty years ago an analytical chemist working in the industry's laboratories used simple glassware, and his single most expensive piece of equipment cost less than £500. Today, technology has given him powerful new tools such as the mass spectrometer, costing £50,000 or more. And it is getting harder to find new areas for innovation. The rate of introduction of new drugs has been decreasing for several years, reflecting the fact that some of the formerly entrenched diseases have been successfully attacked with sulphonamides and antibiotics, and now the pharmaceutical companies have to look further afield to many new compounds in the search for new agents. The industry typically spends about 7 per cent of its turnover on R & D, and has a high proportion —around 15 per cent—of its employees in the laboratories. As a result of this continuing strenuous search half of the most widely prescribed medicines have been developed since 1960. Once a new medicine is introduced and used regularly, its price generally falls. Penicillin, which was an expensive rarity only twenty-five years ago, has become a low-cost medicine available to practically all who need it.

In spite of the careful testing, occasionally a drug slips through whose harmful side effects have not been discovered in time. The case of thalidomide is probably the most tragic one in recent times. The administration of this drug to pregnant women as a tranquillizer resulted in an estimated 6000 deformed babies being born in West Germany, and at least 500 in Britain. The episode shows that great care must be exercised in introducing exotic agents into the body. Every drug must interfere with the biological processes: otherwise it could not have any effect, desired or otherwise. The effects may be slight or strong, depending on the individual who takes it. The more powerful the drug, the greater the degree of caution required.

One case of a possibly over-enthusiastic use of drugs is the prescription of the latest wide-spectrum antibiotics for any minor ailment that could quite readily be cleared up with a less powerful compound. Micro-organisms can become resistant to drugs, especially when exposed to concentrations that are not high enough to kill them. Worse, in certain circumstances one micro-organism can pass on its resistance to another. It can mean that when the drug is really needed to combat a germ that is unassailable by weaker formulations, its power may have been entirely vitiated because the germ has built up resistance to it. There may also be repercussions from the practice of adding antibiotics to animal feeds to prevent disease, foster growth and relieve stress in the animals, particularly those kept in 'factory farming' conditions. Bacteria in the animals are thus exposed to low levels of antibiotics, which may not be enough to kill them, over long periods. This is likely to encourage the proliferation of resistant strains that may be passed on to Man through eating the animals' meat or in other ways. When the antibiotic concerned is used to treat the disease caused by these bacteria in humans, it is therefore powerless.

Thanks to the extensive investigations and operations of the pharmaceuticals industry, diseases caused by many micro-organisms, worms and other bodily invaders, and by food deficiencies and disturbances of the body's delicately-balanced chemical supplies, are well known, as are the treatments for them. There are drugs to regulate the body's responses to sensations such as pain; to stimulate or tranquillize; to regulate blood pressure, heart rate, digestion, etc. Now the battle is being extended to other fronts on which the enemy is more elusive, such as those diseases caused by viruses that are unaffected by antibiotics. In time, the industry may produce many more drugs to help Man regulate his body and its metabolism like a memory pill to speed learning, and pills to improve dexterity, counter stress and fatigue; non-addictive sensation pills safer than LSD and other mind-bending drugs; and rejuvenating pills to keep him young. All the time, though, Man has to remember that whenever he prepares some compound like this that has no natural counterpart he is entering unknown ground. Accidents such as the thalidomide case are the grim price of progress.

Hazards of Technology

Life in the push-button, mass-production age has its physical drawbacks. In the highly mechanized world, muscles atrophy and bodies sag flabbily through lack of exercise; arteries clog with fatty residues, perhaps from too much fatty food. Eyes are strained over too much fine work, and ears are deafened by noises of ubiquitous machinery. Lungs are assaulted by air pollutants, teeth attacked by acids from sugary foods; stomachs ulcerated by the stress of living, and minds subjected to the pressures of living in a mass society. So far, little medical investigation has been performed on these 'technologically induced' maladies.

Yet they are far from being the only harmful effects of technology. Others are more direct, like the accidents caused by the machines with which all technology's many tasks are performed. Man is not designed to fight machines, and when the two come into conflict it is the man that has to give way. People can be ground exceedingly small in technology's mills. Usually it is more expensive and more troublesome to make a machine that is safe than one that is merely efficient, and so there is no reason—other than a social conscience—for a manufacturer to do so, except where compelled by law.

As technology changes, so do the attendant sources of risk. As technological capability grows and machines become more complex, more powerful and more generally used, the consequences of a breakdown in the system become more serious, the results of a conflict between man and machine more tragic for the man. In the latter half of the nineteenth century the steam boiler was a constant source of risk, and explosions were common in Britain. In the early part of the twentieth there were disastrous dust explosions in such places as mills where vegetable oils were extracted. The modern automatic transfer lines, which are widely installed now in car plants, machine tool shops and elsewhere are potentially hazardous mechanisms. The operator may reach into the machine round the fitted safety guards to free a jammed part, say, not thinking of turning off the power. As soon as the jam is cleared the machine starts up again automatically, and he is trapped. Other new techniques or processes add their peculiar risks. Storage systems in factory and warehouse have tended to store upwards (to save space) instead of outwards, and despite greater automation there are hazards caused by the greater height, and by objects falling on to men below. There are also the dangers of electrocution with the wider use of electric tools, and risks of injury through new techniques like the cleaning of chemical plant with high-pressure water jets.

Modern large-scale processes need large quantities—often thousands of tons—of hazardous materials in, or near, the factory or plant. These materials may be oil and oil products in the case of the oil refinery; acrylonitrile, a raw material for synthetic fibres; liquefied petroleum gas; and new and dangerous compounds such as liquid oxygen, hydrogen, or biologically active detergents. The use of inflammable gases for heating or reducing metal treatment and chemical synthesis causes a number of industrial explosions each year in Britain alone, and in fact flammable liquids pose one of the most widespread industrial risks.

Building sites have always represented an especially thorny problem. With the adoption of more system building, the components are larger and heavier and have to be handled by cranes. A larger component can do more damage than a small one, like a brick. An 8-ton ceiling slab, for example, with only a few centimetres at each end actually resting on the supports, has to be delicately positioned by the crane driver and building team, and if the supports are not quite the right distance apart, it can slip off and cause considerable damage. As it is, one man in fifty employed in the British construction industry dies each year as a result of an accident at work.

Among Britain's 25 million working population, there are estimated to be annually more than 1000 deaths at work, 620,000 injuries involving at least three days' absence from work, and 300,000 cases of industrial disease with, again, three days or more off. Some of the occupational diseases have been stamped out, such as mercury-poisoning-madness among hatters, fossy jaw among phosphorus workers, pneumoconiosis and nystagmus among miners, and so forth. Even so, dermatitis caused by reactive chemicals used in plastics manufacture is the commonest occupational disease in that industry, and a machine-shop worker may contract cancer because an oil-soaked overall is in constant contact with his skin. People who are called upon to do long sessions of fine work, or who have to stay visually alert for long periods of concentration—those making maps, inspecting pills in the drug industry, precision engineers, computer programmers and people making semiconductor devices—can develop unpleasant symptoms according to some reports. They range from discomfort in strong light, blurred vision after working hours and pains in the eyes to nausea and other stomach troubles.

Rapid transport brings its share of hazards. Speed in itself has its drawbacks, but there is also the effect of long, fast journeys on the body. Our 'internal clocks' govern the near-24 hour rhythms of a large number of bodily functions. Our body temperature, for one, falls in the hours before breakfast, then rises sharply until 10 a.m., and then remains high until 10 p.m. before dropping again. Our efficiency on tasks requiring sustained attention, judgement and intellectual effort follow a similar 24-hour curve. When suddenly switched to a different time zone the internal clock takes time to re-set; typically, four to eight days: as when one catches a plane in Europe one morning, arriving after a nine-hour journey in the United States on the afternoon of the same day, or when one starts on nightshift in the factory. In the interim one has to force oneself to work at times when one's temperature and efficiency are low. The effect is well known among air-travellers. It has been suggested that it impairs the judgement of politicians, senior military personnel and businessmen, whose hurried timetables allow them little respite between one conference and the next across the world, and who cannot therefore re-orient themselves at each globe-trotting stop.

Speed brought by advanced transport is inherently dangerous. Man is designed for a pedestrian or swimming existence, and cannot take the shocks of high-speed accidents without damage. All other accidents in this sphere pale into insignificance beside those that occur on the roads. Driving is one of the most difficult tasks that many people are called upon to do. The whole occupation of driving is almost that of a devilish game of chance, where the odds are loaded against the human. It is only by dint of extreme skill that people manage to survive for so long.

Just consider it objectively. What happens on our roads is far more deadly than any medieval joust. Two massive metal contraptions come hurtling at each other, each one breathing fire, containing appreciable amounts of inflammable liquids, and fitted with all sorts of sharp metal edges and bolt heads, and glass that shatters into knife-edged fragments—in which the only means of communication

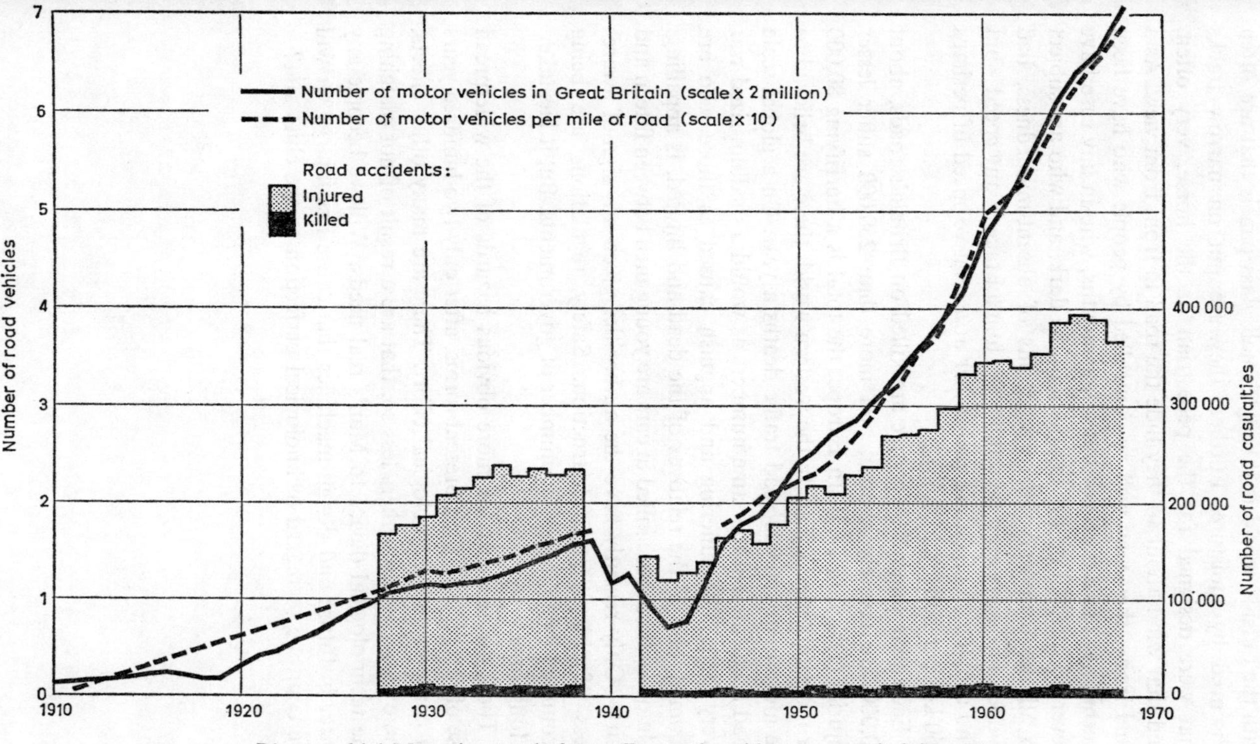

Diagram 14.4 Mounting totals for traffic—and accidents—on Britain's roads

with the outside world is through small glass panels that are often obscured by moisture. Further, they are put on narrow tracks that were designed for the pedestrian or the horse, very often, and that wind about so they hide the road in front from view. As a final irony, they may well be controlled by people who have been learning to drive them for but a few months, who in any case have never been tested on driving them in the dark, and who are subject to all the temporary aberrations, fits of absent-mindedness, bad temper and aggression that afflict the human race at any speed—but who are now confronting each other at a relative speed of perhaps 160 km per hour (100 mph).

Each year some 7000 people are killed on Britain's roads, about 90,000 are severely injured, and more than 250,000 suffer lesser injuries (Diagram 14.4). In Europe, the total is a horrifying 80,000 or more fatalities a year. In the western world, there are believed to be more than 150,000 road traffic deaths a year. On a global scale road accidents kill the same numbers as would a medium-sized war every year. The suffering and anguish caused to those who are maimed, and to the relatives of the dead and injured, is appalling. Almost half of those killed in cars are young men between fifteen and thirty. Only recently have the authorities given any sign of awareness of this wanton destruction. Safety regulations are being promoted by an increasing number of governments. But it has taken a long time.

These are some of the more obvious hazards of the widespread use of machines in their present forms, after only two hundred years of concentrated technological effort. There are many other effects, some more serious, others less so, that are a result of not matching the technological device to Man's real needs. Will we become any wiser in future and design machines that cause less injury? Or will we continue to regard technological perfection as the chief aim?

15 · War: the Crucible of Technology

The New Spirit of War

With typical perversity, Man has put some of his best technological skills into war. No sooner has he conquered a fresh part of the environment than he uses it for military purposes: first land, then air and sea, and now space. Almost since the beginning of his race, when stone axes were wielded against men, technology has been spurred on by a spirit of belligerence.

Yet something has happened to the spirit of warfare in recent times, under the realization that neither side can now win. War was never glorious in spite of what the military historians would have us believe, but at one time it comprised a series of essentially personal battles, with man pitted against man. History's milestones right back into classical times mark the struggles of an individual or a group against the exigencies of fate, or against hostile armies, always with some definite objective. They were invariably presented as being inspired by some noble goal or some great vision. While this view is partly due to an elaborate mythology fostered by the military elite, there is no doubt that at one time success in the physical struggle of war was a mark of accomplishment, a sign of virility, as indeed it still is among primitive tribes.

Until two generations ago wars were comparatively local in effect. It was partly because communications and transport were not advanced enough to keep large armies together and make possible the movement of large numbers of men and supplies around a battle area so that there was sufficient co-ordination for mustering a complete fighting force. Battles involved relatively few fighting men, though casualties increased as weapons were improved and greater numbers took part. At Malplaquet, often regarded as the bloodiest battle of the eighteenth century, allied losses were 23,000. Losses on a larger scale began with Napoleon's campaigns, although disease took many more men among his ranks than did bullet and sword. At Waterloo the British lost 13,000, the Prussians 7000, and the French 37,000. The toll of the American Civil War was far heavier: the dead alone numbered something like 700,000. But with the First World War casualties were for the first time numbered in millions—there were probably more than 7 million killed during those horrifying four years. Mass warfare had arrived. Technology had by then prepared the weapons of mass destruction that are now taken for granted. The Second World War, only twenty-one years later, spurred on technology to an extent never experienced before, improved on these weapons pitilessly, and caused much greater destruction and misery. Since then the world has witnessed the evolution of infinitely more awesome weapons—the nuclear bombs—of such force that one bomb is alone capable of killing a million people.

The mass war of the twentieth century is more impersonal, more calculating; it is a computer war, casualties being coldly measured in units of a million deaths—the megadeath. Whole nations are swamped with a blanket of terror. People no longer matter, individual courage loses its point and its heroism. The soldier of today may not even see his enemy. He may be far from the front line, and may only press a button on a control panel; yet he can release in one missile a destructive force far greater than that commanded by an entire conventional army, and wipe out a whole city at one blow.

The United States and the Soviet Union could annihilate each other as viable civilizations in a day with atomic weapons, or perhaps in an hour. Each can kill 120 million of the other's population immediately, and cause many more deaths subsequently through fallout, fire, disease and starvation. Besides that, more than three-quarters of each nation's productive capacity—and hence a large part of its technology—would be destroyed. There would be no opting out for other nations, no place for them to hide from the rain of radioactive terror. A full-scale conflict would put an end to all Man's attempts to build up his civilization and control the environment.

In such circumstances all-out conflict becomes an impossible course. There could no longer be any point to such a war because all parties would suffer in an equally catastrophic manner. If war ever had any justification, it has now lost it. Nonetheless, the world continues to act as though it were preparing for it. Twice this century the developed countries of the world have split into two camps, each intent on destroying the other. Now there is a different alignment: the countries of the West (within NATO) and of the East (the Warsaw Pact) and Far East (China) glower at each other across the globe. Technology has changed the situation since the last world war ended. With the unleashing of the virtually unlimited might of nuclear weapons, each of the camps has been prevented from attacking the other (or others) by the restraint of self-preservation. For all their nuclear might, the great powers are unable to use it for fear of precipitating the doom of the entire world.

Yet the reasons for which war is undertaken remain, among them the thrust of different ideologies and the desire to acquire valuable resources. Almost certainly these will intensify as the increasing populations in various countries struggle to obtain enough food, water, materials and living space in their limited world. But it will be impossible to resolve such conflicts through resort to nuclear weapons. It is likely that numbers of 'skirmishes' or undeclared wars will flare up while the nuclear powers look on helplessly or try to use their influence—or failing that, their conventional forces—to bring the parties to see reason.

In the future, as in the past, certain countries may see some advantage in waging war against a weaker neighbour. War provides readily definable national goals, for the population and for technology, that are overriding and politically justifiable. All are united for the common good in the struggle for existence. The battle provides an outlet for aggressive instincts that are latent in everyone to some extent, and diverts people's attention from internal problems. In war, the stimulus of the day-to-day feud with the environment is heightened sharply in the struggle against a more readily identified human foe. If the foe does not exist, propagandists can invent one. However, in the circumstances of the enforced stalemate of nuclear armaments, when neither side can act, these outlets no longer exist. Denied the channel for aggressive emotions that seem part of its character, and that have been necessary for it to have attempted to conquer its environment, the frustrated human race turns in upon itself. Society looks inwards, to easy victims of senseless violence.

Crucible of Technology

In warfare, each of the opposing forces wants to have the technical skill to outwit the enemy. Each wants its weapons to be more accurate, destructive and efficient than those of the enemy; and it wants to devastate the other's territory and obliterate his forces while protecting its own, and ascertain what the other side is doing so as to be in a position to retaliate. To accomplish these aims they are prepared to spend heavily. War is a crucible of technology into which nations pour money, materials and men to conjure up new weapons and their aids, almost irrespective of cost. As the mass-production industries are brought to bear on military needs, economic growth is stimulated: the United States economy, for instance, would probably have been unable to sustain its boom were it not for the needs of the Korean War, the Vietnam War and the ballistic missile programmes.

With the growing sophistication of technology of warfare simple weapons have evolved into weapons systems and interrelated land, sea and air forces under computer control, with complex radar-based detection units to warn them of hostile action. Looking back into history we can identify some of the most revolutionary steps in weaponry: among them are the long-bow, gunpowder, the machine gun, the tank, the fighter aircraft, the submarine, ballistic missiles and the atomic bomb. Each one led to profound changes in tactics and force structure.

It is difficult to see what the next revolutionary step will be, now that Man has the power to unleash what is, to all intents and purposes, unlimited explosive force at any point on the earth. Yet evolution continues. Ancillary work has to be carried out on new materials that are stronger and more resistant to extremes of heat and cold, pressure and shock, radiation and other effects of the war environment. New methods of communication and control have to be devised, using integrated circuits that are cheap, reliable and rugged enough to perform dependably in any conditions. Missile guidance systems are an example. Computers as compact and portable as suitcases can direct forces continuously on an entire battlefield, but all the time they must be made more responsive and capable of handling greater quantities of information and of issuing data more speedily. There are radars to detect enemy aircraft and missiles thousands of kilometres away, and they too must be continually improved to detect new threats.

Electronics techniques are becoming increasingly important to military evolution. At present, about 45 per cent of the cost of a missile system goes on electronics. From the control of a single missile, electronics techniques run through the whole gamut of activities—including the collection and transmitting of data about the other side and marshalling the forces on the battlefield. In the

Vietnam conflict, the United States and South Vietnamese forces relied heavily on radar, telecommunications equipment and computers, which all in turn depend on electronic components in large measure.

On the modern battlefield, information from the front can be passed back to base through rugged, mobile radios carried in a soldier's pack or in the back of a jeep. Messages are electronically scrambled for safety. Information may also be sent back to one of the computerized outposts through data terminals mounted in jeeps. Once the computers have digested the reports from all parts of the war zone, they can quickly print out or otherwise display results showing the course the battle is taking. Electronics plays an important part in detection. Enemy forces can be tracked by simple photography or radar techniques, some of which are sensitive enough to spot a single walking man from an aircraft at 2400 m (8000 ft). Other devices pick them out by means of their infra-red radiation (see page 57), by the scent of telltale compounds such as ammonia and urea around an encampment, or by detecting the motion of radioactive dust stirred up when men and vehicles pass through a pre-seeded area. A rifle-mounted image-intensifier enables an infantryman to pick out a target 250 m (820 ft) away under moonlight in open country. Other devices use radar and an audible signal output, so that humans are detected by the whoosh-whoosh noise of their arms swinging as they walk.

As one side invents new ways of detecting enemy activity, whether it is in the jungle, across the high seas or in the air, the other tries to go one better. Much can be unearthed by intercepting and decoding enemy signals. With military communications networks spanning the globe and even extending—via communications satellites—into space, there are multiple opportunities for interception. It is a constant struggle for each side to invent safer ways of coding their data, and better techniques for cracking the codes of the other. Each is constantly inventing new electronic countermeasures so as to make use of the other's electronic activities—e.g. in order to detect and locate the source of a hostile radar beam so that it can be destroyed, or to alert an aircraft when it is being illuminated by such a beam. Alternatively, there are a whole range of devices that can be employed to jam enemy electronics or deceive it with decoy missiles, with 'chaff' scattered from a missile or aircraft, or by other means.

Another field of warfare—it might even prove to be a revolutionary one unless banned by concerted international action—has been opened by the mass production and stockpiling of chemical and biological weapons. In spite of a quite illogical human revulsion from such weapons, which may act much less drastically on humans than explosives the major powers have developed an armoury of such agents. The first chemical irritants (chlorine, phosgene and mustard gas) were discharged during the First World War. During the Second World War large amounts of improved irritants were made, chiefly CN and the more toxic DM. CN has been employed against civilians in Vietnam, and in the Paris demonstrations of 1968. The chief harassing agent made and used in massive quantities is now CS, a harsher replacement for CN. A concentration of one to ten parts per million is enough to drive all but the most determined out of the affected area within a few seconds. It, too, has been used in Vietnam—and in substantial quantities.

Development work has produced gases far more lethal than their predecessors of a generation ago. Indeed, the victim who survives the initial blistering of the eyes, skin and lungs from a modern gas attack may die months later of a breakdown in his bone marrow. There are also the modern nerve gases, developed in the 1930s in Germany as by-products of a search for new insecticides. These gases, now known by such names as GA, GB, GD, the deadliest of all being VX, act on the body's nervous system, produce convulsions, and paralyse the heart and breathing. Death comes within thirty seconds, typically. A single drop of VX on the skin is enough to kill. Still more chemical agents have been produced, including incapacitating gases that induce giddiness and hallucinations, and sometimes maniacal behaviour that may last for days. By the end of the 1960s, the United States alone probably had enough gas stockpiled to kill every living thing on the surface of the earth several times over. Presumably the Soviet Union possessed similar quantities, as it is known to attach great importance to such weapons.

Biological weapons could be the most devastating of all, as they turn against Man some of his malevolent microscopic enemies (see page 291). During the Second World War both Germany and Japan, on one side, and the United States and Britain, on the other, were developing such agents. So virulent are they that 500 grammes of botulinus toxin, for instance, product of a type of bacterium, could kill the entire world population. Other lethal diseases, among them pneumonic plague (a much more deadly variety of the organism responsible for the Black Death in the Middle Ages), pulmonary anthrax and typhus could be spread in this kind of warfare—sprayed from aircraft, fired in artillery shells or rockets, or floated ashore in drifting mines.

By the late 1960s extensive stocks of such weapons had been accumulated by the major powers. However, with the growing realization that this form of attack might just as likely rebound on the attacker, the United States unilaterally renounced the use of biological weapons late in 1969. A further important step came in 1972 with the signing, in London, Moscow, and Washington, of a convention that outlawed biological warfare and required the destruction of all existing stocks of such weapons.

Nevertheless, explosive weapons still form the basis of modern warfare, and much technological effort is expended in finding new explosives of ever greater power and new methods of delivering them in suitable warheads or bombs to outwit the enemy's defences. For centuries, Man had nothing but gunpowder, which was probably provided by the ancient Chinese. Then came fulminate of mercury, discovered in 1800, followed by nitroglycerine and trinitrotoluene, products of chemical technology, later in the nineteenth century, and finally atomic weapons, the result of nuclear technology, in the twentieth. Explosive charges of 'conventional'—that is, chemical—explosives were limited by the practical size of the weapon to the force of some hundred tons of trinitrotoluene (TNT). With nuclear weapons using the unleashed power of the atomic reaction, the devastation that Man could wreak on his environment rose by orders of magnitude, to some tens of thousands of tons of TNT with the (atomic) A-bomb, and tens of millions of tons of TNT with the (hydrogen) H-bomb. While the A-bomb utilizes the fission reaction in uranium or plutonium (see page 146), the H-bomb gets its greater power by going one stage further and using the fission reaction to trigger a fusion reaction in a hydrogen-isotope-containing core.

A dreadful day dawned for mankind on 6th August, 1945, when the first atomic bomb exploded at the Japanese city of Hiroshima, devastating an area of about 1290 hectares (5 square miles) in the city centre and killing about 80,000 people—either at once or within a few days through the after-effects of radiation. It was followed three days later by another which laid waste the city of Nagasaki, killing 40,000 people.

The H-bomb was developed soon afterwards, so that by the early 1950s Man possessed a weapon with an explosive power—in just one bomb—greater than the total of all explosives used throughout the Second World War. If used in warfare, such a bomb would blast out a crater roughly 3 kilometres (2 miles) across and kill practically everyone within about 24 kilometres (15 miles), unless they had taken refuge in underground shelters, and many more further out. In a city like London, about 5 million people would perish, and about 130,000 hectares (500 square miles) would be razed. The associated fire-storm and radioactive cloud would claim many more outside the area of total devastation.

If the 100-megaton H-bomb is feasible, in military terms there would be no point in using that much destructive power since one 20-megaton bomb would be enough to obliterate a city smaller than London. Using a 100-megaton bomb would, in the parlance of strategists, be wasteful overkill.

The horrid preparations for creating bigger explosions have been paralleled in making systems to deliver them on enemy territory. A host of different missiles—intercontinental ballistic missiles (ICBM), intermediate-range ballistic missiles, short-range ballistic missiles, defensive anti-ballistic missiles (ABM), air-to-ground, ground-to-air, air-to-air, and many other types, together with free-falling bombs, and so on—can now be employed to deliver destructive nuclear force virtually anywhere on earth. In response to the wide deployment of ballistic missiles of typically 10,000 kilometre range, both the United States and the Soviet Union have resorted to defensive chains of ABM systems. In such systems small missiles that can get off the ground very rapidly are sited in armoured 'foxholes' beneath massive steel and concrete shields which in some cases may even be proof against a direct hit from a nuclear missile. An array of such small units is deployed behind a far-flung chain of radar stations that detect hostile objects—missiles or aircraft—while they are still as much as 2400 km (1500 miles) away, and follow their course. Once the course and speed have been determined, by feeding the inclination and range data from the radars into computers, the defensive flight of missiles are fired towards the intruder. They home on it, perhaps guided by infra-red sensors that pick up the heat of the rocket exhausts, and explode—destroying, it is hoped, the

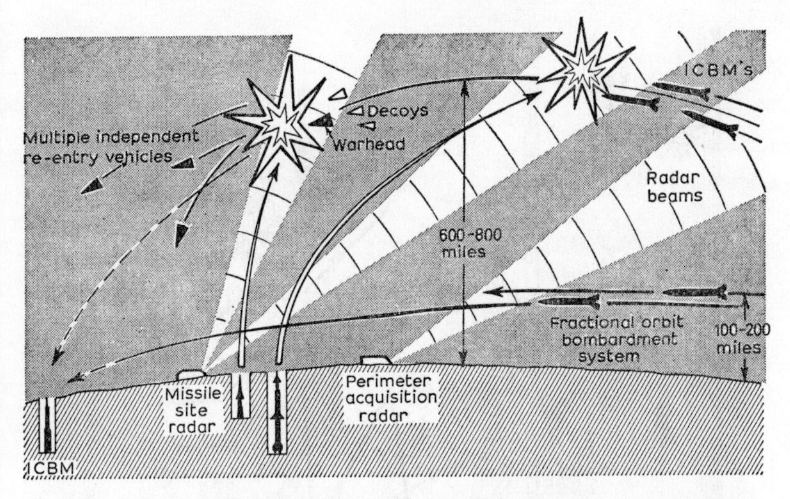

Diagram 15.1 A typical two-stage ABM system, dealing with an attack from ICBMs, with its variants of MIRV and FOBS

intruder as well. In practice, the ABM would probably have two umbrellas as shown in Diagram 15.1 with two sets of radars and missiles, one for long-range operations, which would attempt to destroy missiles while they were still in flight above the atmosphere, and also a back-up line to tackle any that get through in the atmosphere. More advanced ABM systems have been proposed, including seaborne and even airborne chains that would intercept and destroy ICBMs in the early stages of their flight. It has been estimated that the cost of the most sophisticated ABM system might be of the order of $40,000 million. Even the superpowers have recoiled from putting that kind of money into a project bringing no foreseeable return, but nevertheless both are installing limited ABM defences. No doubt China will do likewise when the technology is ready.

In 1968 NATO rejected a system of this kind for Europe on the grounds of cost. However, Britain and other European countries, under NATO's aegis, are co-operating to deploy a European warning system which can electronically call up all-weather interceptor aircraft or surface-to-air missiles to intercept and destroy attacking aircraft. The diagram (page 310) shows the main elements of the system.

Now both sides are devising weapons to outwit the other side's ABMs. One is MIRV—multiple independently targetable re-entry

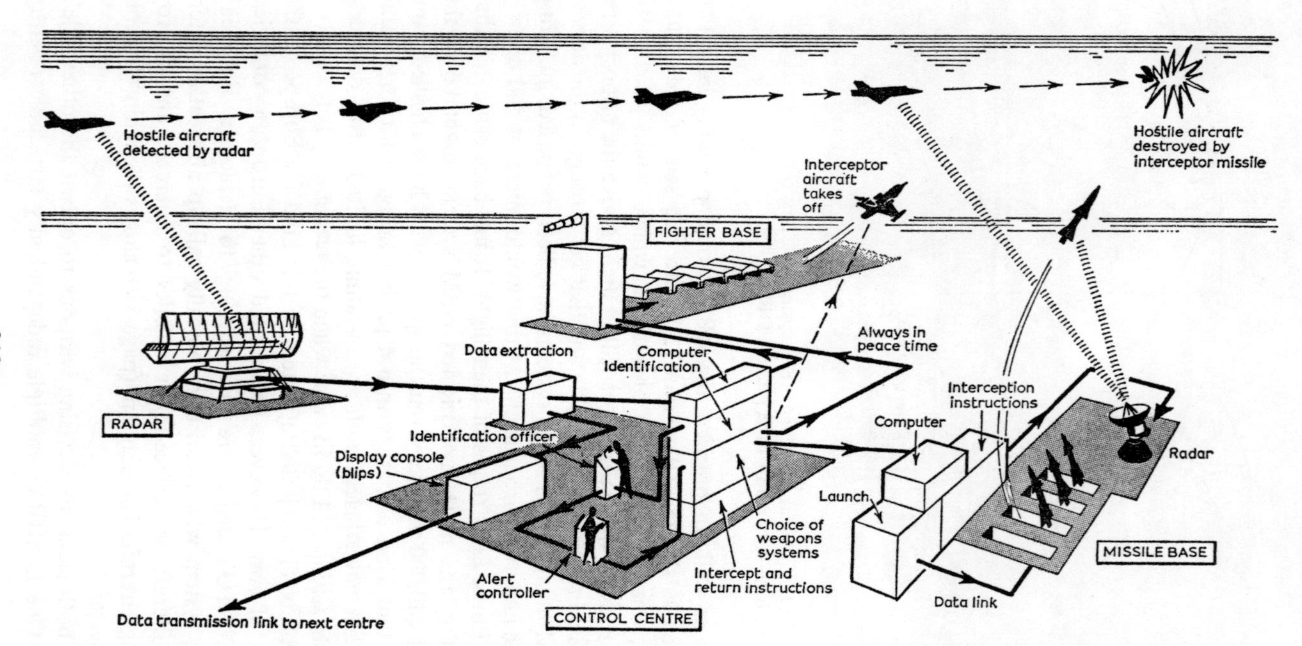

Diagram 15.2 The NATO radar detection chain and associated systems

vehicle—which is basically a scatter bomb. As each warhead capsule comes streaking in towards its target area a booster inside it projects a number of warheads that scatter towards separate targets. The latest idea is to enable each one to guide itself to its target, possibly by geographical sensor, or else by a laser and microwave radar—or perhaps by homing on electronic signals transmitted by the enemy to jam the incoming warhead. Another is FOBS—fractional orbit bombardment system—in which a satellite ejects a re-entry weapon towards the end of its first orbit of the earth. The weapon plunges down to its target from about 160 km (100 miles) above.

These and other devices have well and truly extended Man's warfare into space. There are spy satellites which can take extremely detailed photographs as they float over enemy territory, relaying back to their own side information to aim ICBMs more accurately. Other satellites may be armed with explosive devices of some sort that enable them to seek out and 'kill' hostile satellites in orbit. Between 1957 and 1969, for instance, it is believed that the United States carried out 284 launchings of satellites primarily for military purposes, compared with 174 primarily with civil intent, and the Soviet Union 162 and 170 respectively.

There is another dangerous feature of space activity. A great deal of debris is now orbiting in space (see page 205), and from time to time bits plunge back into the atmosphere, burning like shooting stars. It requires much careful sifting and evaluation, on computers, to make sure that one of these returning bits is not mistaken as a hostile missile, or aircraft, as we have seen. A mistake could precipitate at best an international crisis, and at worst an all-out missile bombardment. Computer centres throughout the world keep check on orbits and predicted re-entry dates and the paths of such space debris, informing the military forces in order to prevent such an incident.

While there is activity in outer space, there is more in 'inner space'—the oceans—in which fleets of nuclear-powered submarines armed with nuclear weapons continually prowl, playing deadly games of hide and seek among the barely charted obstacles of the deeps. With land-based weapons more or less at a deadlock, military attention is turning once again to the sea, both to surface craft and submarines. Beneath the waves, submarine fleets are being armed with MIRV missiles. Above, aircraft carrier fleets are being provided with faster, more manoeuvrable and better armed aircraft to sharpen the attack, while more anti-aircraft and anti-submarine frigates are added to provide cover and a sort of offensive defence.

During the 1970s, the character of the military competition between the United States and the Soviet Union will alter, as the duel in numbers becomes a duel of quality. It seems likely that subsonic bombers will be replaced by supersonic ones armed with both attack and decoy missiles. A new generation of ICBMs will appear—lighter, and of greater mobility, so that they can be moved about by road and rail to offer a less easy target. They will have improved accuracy, better penetration aids and true MIRV warheads which will hit their targets with twenty times the accuracy of current versions. Beneath the waves, quieter submarines will glide at greater depths to make them less vulnerable to attack, while a new nuclear submarine-missile with a range of around 9650 km (6000 miles) will be deployed. On land, a new high-acceleration ABM will be developed to protect sites against enemy missile attack. Both aircraft and submarines will be provided with ABMs so that they can counter missiles from other aircraft and submarines respectively, while land-based ICBMs will be brought into the ABM role. This much seems likely from indications on the American side, and guesses about Russian intentions.

Proliferation and the Balance of Terror

The arms race continues. The world is devoting to military purposes about twice as large a proportion of its output as it did before the First World War—around 7 per cent as against 3½ per cent. In real terms, allowing for inflation this probably represents a tenfold increase. If the trend continued, it would mean a doubling every fifteen years. While military spending in the developing countries is low relative to the rest of the world, it has risen faster than elsewhere—and nowhere more rapidly than in the Middle East and Africa. The diffusion of new technological weapons continues as the arms race infects the developing world. Long-range surface-to-air missiles, unknown in these countries in the late 1950s, have been introduced to around twenty of them. Supersonic aircraft, which were not deployed in these countries in the mid 1950s, are to be found in more than thirty of them today.

Weapons technology can be a lucrative commodity for industrial nations. Whereas before 1955 only Britain and the United States were willing and able to sell sophisticated military equipment, by the mid 1960s there were no fewer than twelve industrial countries in the arms trade, a trade that seeks both economic and political gains. In the twenty years between the end of the Second World War in 1945 and the mid 1960s, it is estimated that over 4500 jet aircraft, about 5000 tanks, 225 warships and numerous guided missiles and small arms were sold or given to the developing countries by the industrial powers. A large percentage were new and of high quality. The commercial value of the arms market in the developing countries over the decade from the mid 1960s to the mid 1970s is put at something like $10,000 million.

Even that is far less than the amount expended by the United States and the Soviet Union combined on their nuclear forces since the end of the Second World War. This sum is thought to lie in the region of $1 million million. The costs of becoming and staying a front-rank nuclear power are prohibitive for all but a few. Technological factors are the most important. Because of the greater complexity of today's weapons systems, they are much more costly and take years to develop. It takes roughly eight years to design a new military plane, for example, from drawing board to maiden flight, and about the same for a new missile. Since technology is continually being pushed to the limit in order to outdo opponents, mistakes are frequent and cost escalation more or less inevitable. Such escalations on American defence projects commonly averaged about 200 per cent in the 1950s, but they were cut to around 100 per cent in the 1960s, thanks to tighter budgetary control procedures. In Britain, the situation is similar. To take a bad case, the Bloodhound ground-to-air missile authorized by the government in 1950 at an estimated development cost of £1·5 million eventually proved to cost £32 million.

Over the years the American military-industrial complex, which is responsible for conceiving, designing and manufacturing weapons, and spans a broad sector of American life from the military headquarters to its network of contractor firms and research establishments, has grown into a business operation with a turnover at the end of the 1960s of $80,000 million a year, consuming about one-tenth of the United States gross national product. At that time, the United States military organization was employing one in every ten working Americans, either in the armed forces or in their suppliers. Comparable figures for the Soviet Union must be as staggering.

In terms of the balance of power, these two superpowers hold the decisive positions. The table estimates their relative strengths in early 1971.

Table 15.1

The Balance of Terror, 1st January, 1971

United States		*Soviet Union*	
Weapons	*Number*	*Weapons*	*Number*
ICBMs	1054	ICBMs	1400 plus
Bombers	1640	Bombers	250
Polaris missiles	656	Polaris-type missiles	208
Carrier aircraft	600	Other submarine missiles	420
Medium-range missiles	64	Medium-range missiles	700
Tactical fighters	400	Medium-range aircraft	700

Despite the cost of becoming a member of the nuclear club, the number of members is growing at the rate of one every five years or so. The United States started off in 1945, was joined by the Soviet Union in 1949, Britain in 1952, France in 1960, and China in 1964. Since the decision to build a bomb was taken by China (probably in 1958), development has been steady. Although the country's internal upheavals since then have slowed progress, no one seriously doubts that by 1980 or thereabouts the Chinese will be able to deploy ICBMs. But no one can be at all sure how they will be used.

In spite of the international treaties on nuclear non-proliferation. and banning nuclear weapons from the seabed, it appears that more members will force an entry into the club before long. The main reason is that plutonium is becoming more widely available, with the growth of nuclear power production. A nuclear reactor fuelled with uranium-235 produces plutonium as by-product as the fuel burns, and this plutonium can be separated at a chemical processing works. The opportunities for diverting some of the plutonium to military purposes multiply every month as more nuclear power stations come into operation. At the moment, more than twenty countries have civil nuclear power reactors and are accumulating plutonium. By 1980, nuclear power stations will be producing, on present estimates, 70 tons of plutonium a year—about one-third in countries that do not now possess nuclear weapons—enough for a hundred bombs of minimum size every week.

There are other routes to the nuclear club. The gas centrifuge process (see page 137) for the production of enriched uranium—another nuclear 'explosive' substance—is one. This separates uranium isotopes in spinning cylinders instead of through the static membranes of the conventional diffusion process, and offers a relatively cheap method of obtaining the most important parts of atomic weapons.

There is also the more remote—but still real—possibility that other nations besides the present nuclear club members will find a method of triggering H-bomb explosions without recourse to fissile material. The discharge of a very powerful laser, or an electrical discharge, are possible alternatives. Obviously, though, the more nuclear materials there are in the world's stockpiles or chemical plants, or wherever it may be, the greater the chance of some of them being diverted for military ends.

As the arms race continues and the world situation becomes if anything more unstable, the danger of world calamity being precipitated by a new, irresponsible nuclear power is very real.

16 · People and Technology

The Mass Society

We live in a world dominated by mass activities: mass production and mass obsolescence; mass planning and mass population; mass creation, mass destruction. Our opinions are moulded by mass media, our leisure hours filled by mass entertainment, our lives dictated more and more by the machinations of big business and huge government bureaux in the name of efficiency rather than humanity. We are becoming a mass society ruled by technology.

We should not underestimate this factor and the effect it has on people. From the very beginning of life men were organized in small social groups. Because communications were poor they did not know what was going on outside their parish or town. Communications now are very much better, and mass transport, coupled with a higher standard of living in the industrialized countries, encourages travel for business and pleasure. It all brings mounting numbers of people into contact with others in faraway lands that they would previously never have seen. For the first time in history, everyone is reasonably well informed about the life, work and conditions of his fellows on all sides. The struggle against the environment still goes on, and the individual is no less vulnerable than when life began. But experience has taught us that the individual cannot accomplish anything worthwhile entirely by himself, and that it is better to work in groups. Mass communication allows everyone to be organized together, and technology permits these larger groups to make a stronger impact on their surroundings.

The personal struggle has become a collective struggle in which the mass of people wreak massive changes on their surroundings by acting co-operatively. In the machine-oriented society that technology has created, collective effort replaces personal effort, collective power takes over from personal power, collective responsibility from personal responsibility. We delegate many of our duties and our privileges to the system. No one can be self-sufficient in the technological society: the individual is significant only as a part of the mass system. By encouraging a technical specialization that relies on machines for its existence, technology makes the individual helpless without the skills of others to complement his own. As society in general becomes increasingly specialized, however, each sector knows less and less about the details of what others are doing. Instead of the individual caring about other people's needs and wants, society deputes public servants to minister to them. It pays specialists to handle its social problems as it pays others to handle technical ones—doctors and nurses are charged with looking after its sick and aged, teachers with imparting knowledge to its young, police forces with keeping its peace, social workers with coping with its misfits, and so on. Society is wholly dependent on them for fulfilling its collective obligations to its people.

Within the system, the individual has greater rights, powers and comforts than he could ever attain alone. If he renounces it, however, or is expelled from it, these are denied him. In order to stay within its ambit, he has to surrender the direction of his life in a growing number of ways to its collective organizations—the huge colossi of power that society has constructed to interpose between individuals and the environment. The chief ones are the 'state' and its departments, and the business organization. To a large extent they determine the environment in which people are born, grow up, mature and die. The individual cedes his freedom to the common will—as expressed through these organizations—and loses his identity within them. Paradoxically, then, if technology confers greater powers on people in the mass, it subjects the individual to mounting restriction. As patterns of population become denser, and activity more frenzied, the consequences of non-conformity become more serious. For example, in the pioneering days of flying, the only restrictions were technical. It was a question of how long one's fuel would last; how long one's engines would run without breaking down; whether the navigation system would work properly; and where the pilot could find landing places. People applauded the aviator's skill and daring. Since improved technology has largely removed these factors, restrictions of another kind have had to be imposed. Now the skies are crowded. Planes take off from major airports with only a few minutes between them, under elaborate rules, and then have to fly in clearly defined 'lanes' according to a pre-arranged flight plan to avoid collision. If a plane strays out of its pre-set and carefully controlled pattern, it could bring disaster. The same applies to road transport and many other fields.

Those who spend their lives in cities—an increasing proportion of the world's population—have to accept many curbs on their freedom. Technology can make life more pleasant in certain ways within the city, by removing the more wearisome chores and troublesome aspects of rural life, and by providing steady employment and services that are unavailable in rural areas. But there are penalties. City dwellers will seldom be able to find peace, quiet, or fresh air. The city may become, for many, a lonely place that grinds away in complete indifference to their individual existence.

With increasing frequency, men have to accept the machine as a part—and a determining part—of their lives. It is partly due to the vast numbers of people involved: mass needs can only be supplied by machine-based mass production, and mass activities have to be regulated by the machines of collective organizations. Because, for instance, there are so many people to look after—in the way of employment, recording of information on taxes, welfare payments, housing, bank transactions, and so on—the traditional state systems founded on the civil servant's pen and memoranda cannot cope and have to be supplemented by machine-based systems. The machine systems, however, are impersonal, cold, rational, inhuman, incapable of seeing the human implications of an action. They can merely operate with standardized, mass information. They have no room for human idiosyncracies. Human identities have to become numbers on a list.

While his dependence on machines, and on the specialists who tend them, has increased, the ordinary man's understanding of them has decreased. This puts him in an odd, frustrating position. It is uncomfortable to encounter mystery. When Man does come upon mystery he either wants to probe and debunk it, or worship it. Technology is mysterious, powerful, and ambivalent. It thus provokes both responses, distrust and awe. Man is naturally suspicious of something inanimate which can do things quicker, or better, than himself, or which can perform tasks beyond his own capability. The person who happily disclaims all knowledge of what goes on under a car's bonnet expresses one aspect of this feeling. The mechanic who treats a machine as an old friend, talking of it as 'she', learning its idiosyncracies, coaxing the maximum performance from it is trying somehow to get round the contradiction.

Until the Second World War, mechanical engineering set the technological pace, and if it was difficult to understand, at least it was visible. People could watch wheels turning and cranks gliding back and forth, and appreciate, even if only vaguely, how the machine worked. Now nuclear, electronic and chemical technologies, and materials studies, have taken pride of place. In these fields technology is concealed. No one can look at an electronic circuit, a nuclear reactor, a chemical plant in the same way and see its innermost workings. The reasons for what is happening can only be sought in the deepest mysteries of the natural world: the flow of electrons under the influence of an electric field, the fission of atomic nuclei, and so on. It is an imaginary world. No one will ever see an electron. The presence of such 'particles' is a convenient but fictional picture for scientists and engineers to work with. How can anyone become capable of thinking of picoseconds as computer designers can, or of light years as astronomers can? The human comprehension—and especially that of the man in the street—wilts before such concepts. Since technology has harnessed the powers of the electron and the atom (through electronics and nuclear power technologies) a growing part of the technological realm, in which these properties hold sway, is barred to most people. As a result, few people are interested in how things work: it is enough that they do. The tiny transistor radio is taken for granted, except when it goes wrong and has to be delivered to a specialist for repair. Much the same applies to cars, vacuum cleaners, television sets and all the other machines that people have brought into use to make their lives easier and more comfortable.

In recent times, the awe with which technology used to be viewed has to an increasing extent given way to suspicion. Alienation from technology has intensified chiefly because the human race has had to accustom itself to unprecedented technological activity.

Until approximately the turn of the last century a man could have been exchanged with a man from an earlier century and he would not have felt all that much out of place. Even in 1800 most people still rode horses for transport, cultivated the land for their living, and knew little of the world outside their parish. Even someone living in Britain in the 1870s and 1880s had these kinds of links with the old, stable, ordered world. The 'time traveller' from an earlier age would have been perhaps 70 per cent on familiar ground. Now such a transfer would be impossible. Technology has cut us off from the past.

It is easier for children. They are growing up amid supersonic aircraft, computers, moon flights, television, nuclear power and the rest. They have access to cars, radios, gramophones, tape recorders, contraceptives, cosmetics, plane flights to faraway places and a whole host of possessions and opportunities denied to their parents. They have more secure and better-paid jobs in better working conditions. They earn and spend more. Financially, they are much more independent, and they are wooed by the mass-production industries that are attracted by their massive spending power. They are becoming more numerous and so more of a power to be reckoned with. In 1970, 40 per cent of the British population and nearly 50 per cent of the United States population were under twenty-five. In the developing world the percentage was even higher. By the year 2000 half the population of the world will be under fifteen. Today's young people everywhere share a certain sophistication when it comes to the material world and technical matters. Besides that, they are maturing younger, and growing taller and heavier, thanks probably to technology's raising of living standards. Thus girls in the United States and Europe today reach puberty at about thirteen, compared with sixteen or seventeen in the mid nineteenth century. A five-year-old boy is, on average, 50 millimetres (2 inches) taller than his counterpart at the turn of the century, and an eleven-year-old is probably 100 mm (4 inches) taller and 9 kg (20 lb) heavier than sixty years ago. The disparity in technological experience and the growing independence of young people as a result of this earlier maturity and greater spending power, combined with an education that stresses the value of personal verification of truths rather than blind acceptance, aggravates the tensions of adolescence and weakens the traditional fabric of social structures, including the family unit and marriage.

In a world in which they find themselves hemmed in by machines and inanimate objects, growing numbers of people rebel, rejecting the mores of a society that lives by technology alone, and seek to regain the standards of primeval innocence, of a pristine, pastoral world unsmirched by technology. The feeling is expressed, more or less clearly, in various 'escapist' activities like 'hippy' movements, the fantasy of the pop-music world, drug-taking and superstitious and magical rites. For centuries the American Indians found solace in mescaline extracted from the peyote cactus as a panacea and hallucinogen. The East has opium, hashish, betel nuts and other things to turn to, while the industrial nations have their tobacco and alcohol, supplemented by imported habits from elsewhere, which they seem to need to reconcile themselves to the unpleasant features of technological society. Drug-taking is spreading in the industrial countries, apparently; a habit first of the rebellious young, it is extending to the middle class. In Britain, the number of registered heroin addicts is increasing at 50 per cent a year. An estimated 12 million Americans have tried marijuana, while the United States contains at least 60,000 heroin addicts. The number is rising.

The 1950s and 1960s have seen unprecedented progress in technology, and unprecedented social upheaval. The natural tendency is to link the two and look to technology's influence for the cause of the strife. It may not be incorrect to do this, for the reasons we have outlined. It has been suggested also that there may be a link between the rate of a nation's economic growth and the level of national arousal, as measured by such things as suicides, road deaths, psychosis (schizophrenia) rate, and the amount of eating and, incidentally, alcoholism. In Britain, with its chronically low growth rate, all the indicators point to a low level of arousal, compared with other less placid and economically more successful nations, who may after all pay a social penalty for technical success. It is an interesting theory. The adjustments that people have to make to accustom themselves to the rapidly changing world are something of a strain; for some the strains are severe. Moreover, there are bound to be frustrations caused by the clash between the individual's desire for more freedom of action, exercising his technological powers to the full, and the increasing need of society to restrict him.

Work and Leisure

In the industrialized countries, where there is ample technology and machinery to back up a worker's brawn, the time spent on work is declining. It is plain that a man has a better chance of making more items with a machine to help him than with his own muscle and a few hand tools. As a result of his higher productivity, he should have the opportunity of seizing more time for leisure. The average man's working week has fallen from eighty to forty hours in the last 150 years or so. Some predict that by the year 2000 the working man in the United States, for instance, is likely to divide his year between 147 working days and 218 days off. The working day would be seven and a half hours, the working week four days long, and the man would be entitled to ten days' statutory holiday plus his annual thirteen weeks' holiday. No doubt other advanced countries will follow the same path, as technology provides the opportunity.

And yet, that assumption needs to be qualified. People can choose either to have more leisure or to stay at work doing overtime and earn more. In the United States, people have in the main chosen the latter. Actual hours of work have hardly changed at all since the First World War. They are currently about 40 hours a week, and on present trends would be between 34 and 38 hours a week by the end of the century. It could be argued that actual leisure has decreased, because journeys to work have lengthened with the building of far-flung suburbs. Working wives leave more housework in the evening for their husbands to do. And more Americans have taken up a second job—doing what is known as 'moonlighting'.

In Britain, the average hours worked in 1958 were the same as in 1938, and were almost the same in 1967 as in 1947, in spite of all the technological change brought in during, and after, the war. In 1964, it was estimated that one worker in six had a second job, and recent estimates suggest that this sort of work is on the increase. (It is fair to say that Britain is untypical in that people work more overtime than in any other European country except France.)

What does appear to be happening, though, is that people's expectations for leisure in the industrial countries are rising. They can afford more household gadgets that enable them to do the household chores quicker and more easily: they can buy a cooker with an automatic timer to turn it on at pre-set times, so that the housewife does not have to stay at home to see to meals herself: they can get their refrigerator, dishwasher, vacuum cleaner, food mixer, toaster, steam-electric iron, electric can opener and a whole host of other gadgets—which may be joined by the 'robot housemaid' within the next decade. The house may be fitted with central heating and air conditioning. Almost certainly it will boast a radio, record player (perhaps a hi-fi set), a television set and maybe a tape recorder. There are drawbacks. One of them is that as living standards rise and more gadgets find their way into the home, children are more likely to withdraw from the family circle at an earlier age, demanding their own room, record player, telephone, radio and television, etc., and creating a private world of their own, electronically isolated from the rest of the family.

More families will be able to afford a second car, and take their holidays abroad together. They will be prepared to spend more on their holidays and hobbies. The mass media, particularly television, have helped awaken appetites for new and more specialized interests. With greater mobility afforded by car and aircraft, people travel greater distances to engage in those hobbies. For example, ski-ing, once the preserve of the rich few, now claims a much wider circle of enthusiasts. Improved communications have made sports like football, golf, and tennis international for both players and spectators, though it seems inevitable that leisure activities will become more mechanized. All games have their special tools, to which technology brings new materials and better performance—plastics boats, steel tennis racquets, even carbon-fibre golf clubs—but as time goes by it creates new opportunities for mechanized pursuits for many more people—e.g. motor-car racing; riding on self-propelled ski scooters; underwater exploration; parachuting and sky-diving; and so on. In the home, do-it-yourself hobbies will be stimulated by tools of clever design and enhanced utility.

However, the most important effect of technology on Man in his relation to work is that it is gradually removing the need to work for large sections of the population. In the future, the world's necessities will be supplied by a smaller proportion of the working population because of increasing mechanization and productivity, and for the remainder it may be a choice of either going to work or being supported by the state in leisure. A separation of incomes from work has already started with old-age pensions, family allowances, unemployment compensation, and increasing investment and property ownership by working people. If work is no longer a duty that everyone has to perform according to the strict puritan ethic, many may wish to renounce it altogether—though others may have become so conditioned that they have to work simply to fill their time. The opportunities offered by such a situation are immense. So are the potential dangers. Almost certainly there will have to be mass re-education to show people how their leisure can be enjoyed (see page 327).

Workpeople and Technology

While modern technology has removed most of the physical strain of work, it has to a certain degree drained it of interest, inspiration, craftsmanship as well. If the Industrial Revolution lifted much of the physical burden of production from the worker's back, enhancing the power and precision of the human hand, it also broke down the production process into many separable parts. Each task became a simple repetitive job which could be done consistently by a wide variety of people. When many goods are made in this way as they are today, few people see the product in its finished state. The extent of their involvement is one simple task—tightening nuts or welding steel plates, perhaps, for a man, and packing chocolates in boxes for a woman—which they have to do hour upon hour, day after day, even year after year. More and more of their work is being planned for them by superior staff. They can hardly be expected to take a pride in their work under such conditions. For these unskilled and semi-skilled people the technological revolution does not provide the same spread of opportunities as for higher-skilled people. They are subdued—not liberated—by technology. In Japan, employees can give vent to their frustrations by bashing dummies of their supervisors with clubs. In the western world employees seize on 'tea breaks', pretend to be ill so as to take illicit days off, and resort to frequent strikes and walk-outs. Anything to get away from the boring repetition and be recognized as a human being again, not just an ancillary part of a machine.

Now machines are taking on progressively higher tasks. To start with, the tasks were straightforward ones like cutting and hammering. Next came assembly and conveying parts between machines. In time, the jobs of lower management will come to be replaced by computer. Further into the future, the jobs of middle and higher managers will be threatened.

Machines are becoming more complex, more costly, more capable and the man tending them—especially the maintenance man—has to become more skilled and better qualified educationally and professionally so that he can understand them. If he cannot be retrained for the higher-skilled job, his future is bleak, or at least insufferably boring.

We should have had ample warning already of what machines can do when introduced without proper thought. When mechanization began in earnest two hundred years ago in Britain's Industrial Revolution (not long in terms of human experience) the workers were dreadfully, inhumanly exploited. The first widespread unemployment that technology caused was probably among weavers displaced by the combined forces of Sir Richard Arkwright's spinning frame, which out-wove them indefatigably, and Watt's steam engine, which powered the spinning frame and other machines. Prosperity came to the managers and mill owners, while the workers languished in poverty. Industrial towns suddenly mushroomed without amenities or sanitary facilities worth speaking of, although some more enlightened employers did what they could amid the general neglect. As technical progress continued swiftly during the nineteenth century, social conditions by no means kept pace. Workers tried to protest; they were brutally suppressed. Groups of redundant workers, calling themselves Luddites, risked a death sentence when they went round smashing machinery they believed had replaced them. Not for another generation did the benefits of increased productivity seep through to the people whose labours had made it all possible.

Though working conditions and wages were steadily improved, the twentieth century had another traumatic experience in store, the depression of the 1930s. In those grim times Britain was accustomed to 2 million people out of work; and in 1932 the total rose to nearly 3 million, one in every six men being unemployed. Continental Europe suffered as much. In the United States the position was worse than 12 million unemployed, one man in every four. If today conditions as regards redundancy, unemployment, replacement and re-training are much better, there is still far too much of the 'them and us' attitude about labour relations. Too little consultation takes place between management and men before a new step is taken that affects them all, such as bringing a new machine into the factory. Much of the distrust engendered in the bad old days lingers on.

Understandably the trade unions, haunted by memories of the depression, and perhaps possessed with a sense of the history of the Industrial Revolution and the exploitation that followed it, are probably more anxious about the effects of automation than any other single subject, barring wages. It is a central theme of unions that the employment of their members should be safeguarded. However, as it is technology's aim to replace human brawn and skill with machines, there must be conflict. That brings unions into collision with management, the advocates of technological progress, amid an archaic industrial relations structure.

One of the most telling indications of the social organization in this sphere having slipped behind the technology is the frequency of industrial disputes. The last few years have been particularly bad from this point of view: British unions are among the worst in the world in resisting change, partly because their roots are deeper and stronger than elsewhere, and on account of the stability of the British social system and the British character. The rivalry between unions—often between the old craft unions and the new ones for managerial and scientific staffs—the demarcations they impose between different jobs, and the conditions of apprenticeship to which they have until recently held firm, all are becoming increasingly irksome and obstructive to modern technology.

The disputes between unions are nearly as bad as the disputes between unions and management. The organization of labour and management and the relations between them are about a generation out of date in Britain—less so in many other countries—and while suitable for a small, privately-owned family firm indulging in small-scale production by craftsmen, they are quite unsuited to mass production in large factories. Time and time again the conclusion about disputes between employers and employees has been one thing: poor communications. The situation can be improved by having worker representatives on the company board, although then the information that they get has to be passed on to the rank and file, which is no easy matter. In order to make substantial headway, however, the attitudes of both sides will have to change. Whereas it takes a short time to change a machine with the aid of a few tools, it takes a good deal longer—and a great deal of persuasion—to change a man's skills or attitudes.

If the problems of the industrialized countries seem serious as they strive to come to terms with technology, those of the developing world are terrifying. Unemployment, next to hunger, is their grimmest problem. In the developing countries of Latin America, Africa and Asia there are about twice as many people under sixteen as in Western Europe and North America. This means that, in proportion, those countries have to provide twice as many jobs. Their plight will worsen. The increase in the number of people who could work if work were available, was expected to reach 162 million in the developing countries in the decade 1960–70, and 226 million in the following decade. That contrasts with 50 million and 55 million respectively in the industrial nations. Widespread unemployment and under-employment in the developing countries is affecting young people, and better education does not necessarily help. The Asian characteristic of higher unemployment rates among the better educated is beginning to appear in Africa.

Learning to Live with Technology

In a world increasingly affected by technology, education has to try and prepare people to meet and surmount technological problems. Because of rapid technological change, the subjects and methods that a young person learns at college or at work will not necessarily be adequate throughout his or her life. An estimated 150 million jobs will change their character over the next thirty-five years under the impact of advanced technology in Europe, the United States and the Soviet Union, and it will probably be necessary for everyone—working people at the very least—to 'go back to school' in order to learn about the latest devices and methods, particularly in a fast-moving technology such as electronics. The ongoing education process should ensure that a man (or woman) is ready to change his skills several times, to be retrained once, twice or even three times during his working life. This requires a new approach to learning which cannot any longer be thought of as a once-for-all activity undertaken in childhood and, as with most people, abandoned there. People will need to learn how to work; they may also need assistance in learning how to use their leisure time profitably. For example, a boilermaker will attend a government training centre to learn to become a carpenter or a capstan lathe operator; a carpenter may be retrained for work in an electronics factory. The manager will attend a business school to learn about scientific management and computers. The office clerk or secretary may learn to become a computer programmer. Housewives may wish to attend evening classes in hobbies and homecraft subjects, learning first aid, perhaps, or a foreign language or foreign cookery. Evening instruction for adults is already a feature of many communities, helping people to pursue either hobbies or subjects involved with their careers, and complementing the full-time educational systems.

Education is already one of the largest businesses in the industrial world. Britain spends about £2000 million annually—nearly 6 per cent of the gross national product—on it, about the same as on defence, and about twice as much as on research and development. It is more than three times the total of a decade earlier, and it seems that the total and the share of national wealth spent on education will continue to rise. Education has been termed the United States' greatest growth industry. While business and other private investment, and personal consumption spending, have risen by about three times in the past twenty years, spending on education has risen nearly eight times.

Most of the vast expenditure goes in bringing old buildings and facilities up to date, and providing more buildings and equipment, teachers, textbooks, playing fields, and so on: that is, providing more of the same. This, it is true, has had impressive results. In Britain, for instance, some eighteen new universities have been established over the past twenty years either as new foundations or created from existing colleges of advanced technology, which means that the number of universities and of their students roughly doubled in that time.

Yet so far technology has had little effect on methods of teaching, and for reasons of cost and ease of use the blackboard is often still the teacher's best friend. However, in these days of providing equal opportunity for all pupils while there is a chronic shortage of teachers, mass teaching methods are starting to be applied. These are especially relevant where children are taught in huge classes. Closed-circuit television and film projectors allow a whole class of hundreds to see an operation or a physics experiment carried out by a demonstrator that would normally either not be done at all or else only be seen by ten or so students at most. More ambitious audio-visual aids combine pictures and sound in pre-recorded lessons that the teacher can play through and then elaborate on for the class afterwards. Some systems link together tape recorders, or videotape recorders, on which the teacher can pre-record his comments together with sounds from other sources, and slide and film projectors so that the whole show can be replayed without the teacher's intervention. Others amount to local broadcasting sets; a group of apprentices in a noisy factory, for example, each wearing a pair of headphones, can thus hear a pre-recorded tape commentary plus live comments from an instructor relayed to them while they, or a demonstrator, shows them how to do the operation concerned—e.g. using a lathe.

Another approach, programmed learning, was pioneered in the Second World War to train military personnel. It can be administered through printed material in the form of a book, or on film through a teaching machine. In each case the material is presented in a series of steps, imparting pieces of information, with questions to test the candidate's understanding. With the machine, the student gives his answer to the questions posed in the text by pressing buttons that turn the film through the machine at each step. Either it goes on if he is right, or it takes him back again over the step if he is wrong. Such machines allow each pupil to go at his own pace. To some extent they lighten the teacher's load and allow him to perform more creative tasks and attend to each pupil's questions. However, they are expensive, and inflexible. Experience with language training for the Forces in the Second World War emphasized speaking rather than reading knowledge, and improvement in the techniques of magnetic tape recording and programmed instruction made it feasible to devise the 'language laboratory'. In this the students practise speaking foreign languages in individual booths, wearing headphones and listening to a pre-recorded lesson, while the teacher helps each one in turn (Plates 16.1 and 16.2). Such installations are becoming more popular, notably for 'crash courses' for people who want to acquire speaking facility in a foreign language quickly before a visit abroad.

The most ambitious way of applying technology to education is to bring in the computer for what is called computer-aided instruction (CAI). The student works through the programme of instructions sitting at a computer terminal. His answers go to the computer, which has been programmed to say which are right and wrong, and to work through the instruction material at the student's pace. Though these systems are at a fairly early stage, in the best of them it is as though there were a teacher at the other end of the line, so 'human' can the response be. Exciting though the prospect is, CAI is prohibitively expensive for most education establishments at the moment, unless they already possess a computer or a computer link for other purposes. But time-sharing and the trend towards cheaper computers are bringing the costs down. In the United States, pilot schemes have been carried out and it is expected that within the next decade individual CAI will be widely used in elementary schools for reading, spelling and arithmetic.

As the costs of computer systems fall and as more computer-based data banks and information retrieval systems are introduced, the scholar will in theory be able to obtain from a computer terminal the answer to a question in a fraction of a second. In that event, it would become far less important for the pupil to have to memorize factual detail. All he would need to know would be where to find the information. This alone has great implications for the syllabuses under which pupils will be taught in the future, and also for the ways in which they are taught. Instead of concentrating on teaching them masses of facts, the teacher (or the teaching machine) will be able to fill the time during lessons more constructively by emphasizing the principles involved, and their application, giving more illustrative examples—perhaps on slides or film.

In terms of its conception and the employment of technology, Britain's Open University, which started its courses in 1971, is probably the most notable innovation of recent years. It offers new educational opportunities to those who were denied higher education, who missed offered educational opportunities, or who wish to master new technology. The combined use of radio and television lectures, broadcast at times when working people can watch or listen at home, written material and home experiments, audio-visual aids, and individual and group tutorials given in local study centres and residential summer schools mark a radical approach to adult education. Almost inevitably, experiments elsewhere will follow similar lines, particularly the employment of television.

However, educational technology like educational opportunities in general is not equally divided among all countries. While an advanced country is talking about language laboratories and CAI, a poor country cannot find enough teachers to teach its people to read their own language. Fewer than half the people in the developing countries are literate, and less than 10 per cent of the children aged between fourteen and eighteen are at school. There is little chance of them learning about life let alone about technology. Whereas in such countries as Britain more than one in every hundred employed people has at least twelve years' education, in developing countries it is not quite one in a thousand. In some African countries, it is not even one in 10,000.

The pattern of education in the industrial world has been heavily biased towards the physical sciences and the technologies associated with space and defence as a result of the vast expenditures lavished by the United States and the Soviet Union on their space and military programmes (see page 191). It has had its effect on teaching throughout the world. Another result of the heavy investment is that the facilities available for research and teaching—which are traditionally linked in most higher education institutes—are much better in the United States and the Soviet Union than in many other countries. This prompts a flow of manpower to the well-equipped areas—notably North America—in search of better opportunities both in terms of salaries and of research facilities, and a receptiveness to new ideas. The 'brain drain' is no new thing. Scholars have always travelled to places where the best teaching and research are conducted, despite restraining edicts.

Although brainpower is much more important to a nation in modern times than it was in a predominantly rural era—when most wealth existed in fields and flocks—such restrictions are seldom possible. The flows take place across greater distances nowadays, out of developing countries into industrial ones, and from the other western states into North America. In 1966, for example, more than 9500 qualified scientists, engineers and physicians emigrated to the United States, almost 4400 of them from the developing countries. In the decade from 1951 to 1961 Argentina lost 4000 trained people to the United States, including 5 per cent or more of its precious output of engineers, doctors and teachers. The implications of such losses of skilled people are serious for all countries.

Whatever their differences—and there are many—the views of leading educationalists in different countries seem to concur about the need for education to encourage a receptive state of mind in pupils. Education should teach people how to learn so that they can go on learning throughout their lives—as much from mistakes as successes. But modern education based on the spirit of scientific inquiry that first dawned in Europe in the seventeenth century does more than that. It teaches people to experiment, to find out things for themselves, to question accepted values and only take them to heart after personal examination and verification. Now this attitude obviously will not be confined to the classroom; it will spill over into the rest of a man's life. People who are used to questioning accepted values and practices in laboratories, colleges and factories will naturally question values that confront them in the rest of the world. If education teaches children to doubt facts and demand a proof for every statement, it will encourage them to view everything that they encounter with equal scepticism. It will in fact pave the way for continuous social upheaval. By laying such stress on education, society is preparing the way for its own continual destruction and renewal.

17 · Guiding Technology

New Powers and New Values

In purely technical terms, technology has been proved many times over; but it has been used in almost complete disregard of the long-term well-being of the human race and its surroundings. However, in the last few years public attitudes, which must play the major part if there is to be any change in this state of affairs, have altered. Public wonderment at science and technology has given way to hard-headed assessments, which frequently extend to downright scepticism and mistrust.

The possession of new technological powers is forcing Man to question moral and ethical values which have been accepted for nearly as long as the civilization that gave rise to them. The practice of medicine has always been regarded as inherently good because it saves life and alleviates suffering. But when it is possible to transplant organs from one body to another, and prolong life artificially with such devices as the electrically powered respirator, the question arises whether we ought to exercise these powers. Death itself needs to be redefined in the face of improved medical methods that have pushed back the boundary between death and life.

As with death, so with life. A number of the world's most horrible diseases have been eradicated. But as a result the world's population is increasing so rapidly that there is too little food. Present attempts at birth control are on too small a scale; a much more drastic programme is needed and the Catholic Church, for one, would be against it. Is it right, in the meantime, to try and preserve the lives of the aged and the hopelessly infirm, even when they themselves wish to die? Is it right to procreate new lives when children will be condemned to an existence of under-nourishment and poverty? The technological means of birth control, sterilization and euthanasia all exist. We may have to accept all of them before long.

Such issues bring the whole purpose of science and technology into doubt. Technical 'solutions' to our problems in turn often create more problems than they solve. Technology is a two-edged sword that can be wielded to good or bad effect.

The actions that we can now perform with its assistance are so far-reaching that they can change the environment radically. The most obvious case is war. Our balance on the tightrope between annihilation and survival is precarious. News of a small border incident in India or South East Asia, say, is flashed around the world in a few hours. A small, smouldering incident in a faraway place could suddenly flare into all-out global war, in which technology would destroy its creator. There are enough weapons stockpiled to wipe out the civilized world in a matter of hours. More nations are obtaining the means to make atomic bombs and deliver them. Even the testing of such a bomb spreads fallout over the whole earth.

Other dangers exist, too, that are not so obvious as this threat of immediate destruction but no less painful in their own way. I have already mentioned pollution. At a less dramatic level, technology is driving out much of the world's romance and mystery. Destroying old beliefs and the old qualities of life, it offers nothing in exchange but material possessions. There are demonstrably no gods on Mount Olympus, when aeroplanes have shown us its empty heights. There cannot be angels above the earth, or spacecraft would have detected them. One can no longer discern the hand of deity in the thunderbolt: it is just a chance electrical discharge. A good harvest is no longer to be bought by placating the gods: it is obtained by preparing the soil in the right way, planting seed of the proper strength and breeding properties, applying the right amounts of fertilizer, water and pesticide at the correct moments, and so on—a dry, rational sequence of cause and effect. The voice of reason, of scientific exactitude, tells us that it must be so. If we carry intellectual analysis far enough, says the voice of science, we shall find a reason for everything. Science determines what is 'right' and technology applies it. The world is a more predictable and less exciting place.

Science and technology deal in things that can be measured, like kilogrammes of steel or metres of cloth. But it is more difficult to put a figure on the effect that science and technology have on human beings, which is after all their only value. We can but try to quantify the value of living longer, being healthier, eating better, and having comfortable homes, shops, offices and factories to live and work in. On the other side of the balance, we must try to measure the cost of the destruction of peace and quiet by busy machines, the disfigurement of beauty, the cost of pollution of air and water, the injuries inflicted by our machines, the havoc wrought by technology.

Science and technology are no respecters of persons. A machine feels no remorse for the men it replaces. People are usually a nuisance in the technological system.

This we have been slow to recognize. We have allowed our preoccupation with technological power to blind us to the dangers of believing that technology can solve all our problems. It is hard to refuse technological projects since the short-term gains can generally be made to look convincing from the economist's point of view. There are however, no counteracting statistics to show what the losses may be in environmental and human terms. A motorway built through the centre of a city speeds traffic but can make life miserable for those living near it, and causes inconvenience for those who want to cross it. If a supersonic aircraft is a great technical achievement, thousands on the ground have to endure the sonic boom trailed in its wake. Even the humble transistor radio can cause considerable annoyance when used thoughtlessly.

The Turning Point for Technology

The chief duties of society in managing technology are plain enough: to prevent over-population and all its attendant hazards; to keep peace; to supply the world's peoples with food and water and other basic requirements; and to control the effect of Man on his surroundings. Up to now, effort has not matched need in any of these areas.

It is no simple matter using technology properly. Not even the United States, the world's most technologically advanced power, has found the way. Many people all over the world are perturbed about the seemingly inexorable march of technology and are perplexed about how to control or direct it. They include all kinds of people from vociferous students to distinguished government leaders.

One major task will be to limit growth. Past generations have worshipped economic growth and productivity. The whole ethos of the industrial world and all its development have been based on growth. Politics, economics, business strategy accept expansion as the right and proper objective. What national or global goal would we have at present were it not for 'economic growth'? Since this is society's goal it has become technology's, too. In the past, science and technology have been allowed—indeed encouraged—to grow.

The number of scientists in the world has been doubling every twelve years, and the expenditure on research in most western countries had been rising by about 15 per cent a year—that is, doubling every five or six years—up to the beginning of the 1970s. The rate is about five times as fast as the income of Britain or even of the United States is growing, and so it obviously cannot continue indefinitely, if only on economic grounds. The same is true of technology, which is fed by the discoveries of science. Technology has been geared to producing more, consuming more, discarding more in an unthinking whirl of mass consumption and planned obsolescence, urged on by ever more refined and insidious advertising. Both technology and science seem to have a self-sustaining impulse driving them on to ever more ambitious projects.

However, in pursuing unlimited growth and unrestrained consumption, Man has recklessly squandered the world's precious resources, fouled its air and water with wastes and drastically upset the balance of his environment—all for the short-term gain of comparatively small numbers of people. It is a process that cannot continue. The resources are limited and cannot sustain unchecked growth in any activity whether it is the procreation of the human race, or feeding the fires of technology with fuel. Each day 200,000 more people join the world's population needing food and drink, housing, fuel, medical care, education and employment. These things cannot be supplied in limitless quantities. Nor will the reserves of hydrocarbon fuel last for more than a few centuries (see page 144), a very short span in the context of human development. Some irreplaceable materials like tin, mercury, copper and lead are becoming increasingly scarce and expensive as old deposits are worked out. With the aid of technology we can do much better to try and meet human needs than we could do without; but on the other hand we could continue to overrun the world and perish by exhausting our accessible stocks of materials and fuels, and space for our discarded spoils.

Our generation is the first to realize how much it owes to technology, to realize that technological progress cannot continue as it has in the past. It is an important turning point. For the first time in history Man will have to curtail growth—of science, of technology, of population and of power over the environment. Society will have to come to terms with the real situation of a planet of limited resources, and of a delicately balanced environment which is incapable of supporting the limitless expansion of any one of its species. It means taking a fresh look at the functions of science and technology.

The future demands a new economic outlook, in which the constraints are not those imposed by banks, governments and conventional wisdom but by global materials balances, living space, population figures, and intangibles such as 'quality of life'. Stability, not growth, must be the target. Governments must take more forceful action to restrict the right of people to breed freely by levying financial penalties on those with more than a given number of children. The world's leaders must forget the lure of growth and seek to raise the standards of living of the poor majority, while holding back the rich to where they are now; and they should not be merely concerned with material possessions but with health and happiness, too, and the fulfilment of deeper needs than plain acquisitiveness. They must limit industry's growth and allocate resources of all kinds among different kinds of industry, and they must guard against over-consumption and pollution. They must take powers to deploy manpower where it is most needed and to regulate wages, prices and social payments. The cutback will be more severe in the rich, industrial nations because one person there uses up more of the earth's resources of materials and energy in his lifetime than one person in a poor country. Those who have most today will have to forgo most in future.

One of society's main tasks will be to correct this imbalance between the rich nations and the poor (see page 37). For the first time in history, technology enables everyone to know about, and aspire to, the good things of life, but at present the spoils won by technology are unequally divided. To emphasize this point: one-tenth of the world's peoples share 60 per cent of the world's income, leaving precious little to be spread between the other nine-tenths. The rich nations—preoccupied with the internal problem of rebuilding their economies after the Second World War—have only just begun to tackle the task of helping their poor neighbours. While the rich nations find it difficult to subscribe even 1 per cent of their gross national product in aid, the developing countries find it hard to absorb the aid, for a variety of reasons. Money too often finds its way into wasteful prestige projects, or the pockets of government officials.

Not only is aid insufficient in volume; it is frequently misplaced and premature, saddling developing countries with over-ambitious technology that they cannot use to the full. Much of the hydro-electric generating capacity in Africa falls into this category. Often much simpler things are of greater use. Although industrial nations are inclined to equate the value of development aid and its cost, a small, well-administered programme can be just as effective as a large, ill-defined one. The industrial problems of automation, redundancy and unemployment, serious in the developed countries, are no less serious for the developing ones, with the fundamental difference that at the moment the latter need jobs more urgently than automated machinery. With so many of their people unemployed or underemployed it is absurd to introduce highly mechanized plant with elaborate machinery. Straightforward, rugged machinery that is easy for relatively unskilled people to operate and maintain is more appropriate. The prime need is to match the aid project with the needs of the community it is supposed to help. In future, aid programmes must be both more generous and better planned.

Unfortunately, technological standardization tends to make all countries adopt the same technological norms, regardless of each one's indigenous needs, cultures or social organization.

However, the situation is starting to change, irrespective of aid. As wages rise in the rich countries, pushing up the manufacturer's costs, the industries on which these countries built their technological structure—textiles, agriculture, iron and steel making, and shipbuilding among them—are gradually being transferred to the poorer countries where manufacturers can pay their staff lower wages. The mass-production industries, such as cars, plastics, and washing machines, will be the next to go. Moreover, the search for fuels and materials is spreading into the hitherto unexplored or unexploited parts of the world. As the rich nations deplete or exhaust their reserves, the developing countries that possess them in plenty will command a better bargaining position as suppliers of raw materials. This of itself will enforce a new relationship between the two.

Large corporations can be as effective as governments in the worldwide application of technology. Rising standards of living and technological standardization are creating vast new markets for their products, ranging from soft drinks to the Pill, and from cars to sewing machines. The world is becoming one market for big business. If trade knows no frontiers, it is still badly hampered by local obstacles. The growth of multinational companies is making national boundaries—and differences in technical and safety regulations for products, in patent laws, in taxation and so on in each nation—look increasingly archaic and irrelevant.

Multinational corporations may command financial resources larger than some national budgets (see page 120), and it is forecast that within perhaps twenty years no more than 300 huge multinational companies will come to dominate world trade. Irrespective of numbers, their growth is reshaping the world economy.

These corporations must adopt a different approach more in line with the humanitarian view of technology's role. They must work with each other and with governments to a much greater extent, for the good of society instead of the good of their shareholders. Competition must be limited, and that may mean government regulating their profits, prices and rate of innovation. The term 'profit' may have to be redefined to take account of the way a company treats its employees and its value to society. At present, a company expects to recoup the costs it incurs in the innovation needed to launch a new product (see page 27) from the product's sales. The process of innovation can be thought of as a chain linking the laboratory with the shop so that while discoveries change science, innovations change technology after lengthy development. Typically, its stages run like this: pure science (discovery); applied science (finding out the validity, extent and limitations of the original discovery); invention (evolving a usable device or process employing the discovery); development and construction of a prototype; production in the factory or construction of the process plant; marketing of the product; and lastly, sales and profit. Each stage is a hurdle at which many ideas stumble and fall. Diagram 17.1 shows how few new product ideas actually get through this series of hurdles finally to reach the market. Diagram 17.2 shows the typical return a company reaps from a new product as it passes through the customary cycle of growth, maturity and old age.

In the past, many good ideas have not been taken up successfully for a long time because the associated technology was not far enough advanced. The helicopter, for instance, was suggested by Leonardo da Vinci in the fifteenth century and yet could not fly until the twentieth because the necessary compact power source was not invented; nor were the steels and other alloys necessary to build it. Another cass is the communications satellite (see page 200). The use

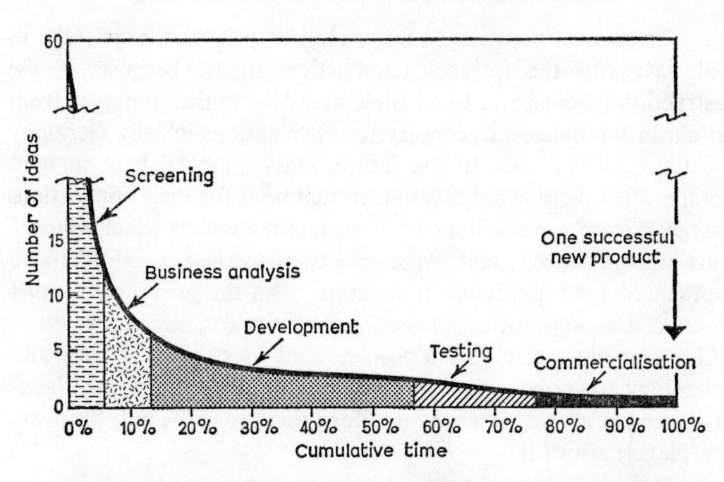

Diagram 17.1 Mortality of new product ideas: it takes almost 60 ideas to yield one successful new product (Booz, Allen and Hamilton Inc.)

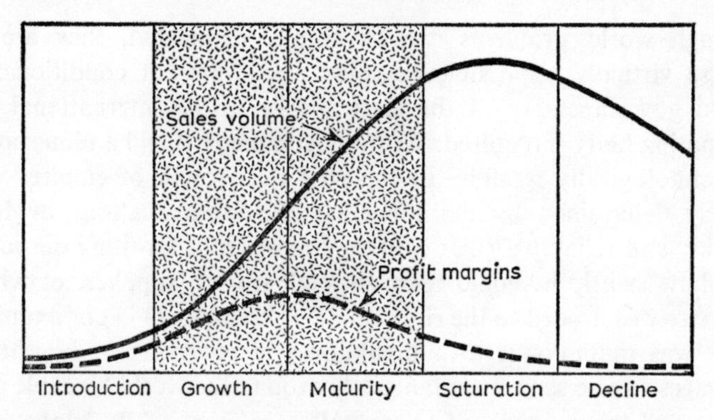

Diagram 17.2 The basic life-cycle of a product (Booz, Allen and Hamilton Inc.)

of artificial satellites to provide global communications was proposed in 1945, but not until the development of the transistor, the solar battery and very sensitive receivers to detect the very weak return signal against the background noise could the idea be put into practice. Another classic case is Babbage's analytical engine, forerunner of the electronic computer (see page 82).

Not only are ideas obstructed because they are ahead of their time; they are often hindered by previous successful ideas. A good innovation can hold up others. Steel in Britain was held back by the heavy investment in wrought iron; electricity by the large stake in coal gas; and the internal combustion engine because of the obstruction from steam. In all these areas the initiative passed from Britain in the nineteenth century to other nations, chiefly Germany and the United States. In the future, ideas must be kept in cold storage when there is not a demonstrated need for their application. That is, governments will have to regulate the rate at which innovations are introduced; and if the society concerned is judged to be unprepared for a particular innovation, then the government must rule that it should not be released in quantity for, say, ten years.

Governments must take a firmer hand in steering science and technology towards socially desirable goals. They must ask themselves, not, 'Can we, as a nation, afford this project?', but, 'Can we, as a planet, afford it?'

The Government of Technology

Though world problems can be simply delineated, they are of course virtually impossible to solve under present conditions. A world government or at the least a very strong international co-ordinating body is required. World government, if still a pious hope, is technologically feasible. In the past the growth of empires was largely determined by the efficiency of communication: by how quickly and reliably a ruler could get news from far-flung outposts, and how swiftly he could then dispatch troops, supplies, or whatever else was needed to the right point. A breakdown in communications was tantamount to a breakdown in government. Now that messages can be sent instantaneously around the world by cable and radio, and troops airlifted to virtually any part of the globe in a matter of hours, it is technically possible for one ruling body to direct world affairs. A central government machine could, in theory, co-ordinate the activities of the world's states for the good of the planet, as far as that is understood. The United Nations is the obvious focal point, although nationalistic squabbles blunt its peacemaking and lack of finance hinders its activities. It does anyhow act as an effective clearing-house for information through its many agencies, and seems to be exerting a growing influence on universal, non-political problems.

Besides effective communications, planners in government as well as planners in industry have modern scientific management techniques to help them today. With the advent of the computer these techniques—among them operational research, discounted cash-flow, preventive maintenance and technological forecasting—are becoming powerful tools that help sharpen the manager's acuity in his three basic tasks: defining objectives; working out strategies; and arranging for implementation and control. In particular, systems analysis should prove a valuable aid since it attempts a synthesis or integration of approach. It implies thinking of a problem in terms of the complete system involved, instead of the workings of isolated parts, which is just what has been lacking before in our treatment of technology.

In the technological society individuals will also be subject to ever stricter control by government acting for the rest of the society. There are too many people packed together in many parts of the world demanding too much space, food, water and material goods—and all of them expecting technological freedom—for there to be much room for further growth, or for unfettered independence for everyone. In order to make any appreciable impact upon the world's problems, each individual must feel that he himself is involved. Everyone must feel responsible for determining technology's future, which is going to determine the future of the planet. One generation may not think it amiss to take tiny nibbles at the countryside, to scoop out ores, coal and clay, to dam untamed rivers or fell virgin forests, and to pollute air and water; but one generation's lapses can easily become the next's curses. While it is easy to protest about the injustice of one person to another or condemn someone for a specific crime, in the mass society there is no way of either identifying the people responsible for its crimes, or of redressing them through fines, imprisonment or some other punishment. It is as yet impossible to impeach the human race for crimes against itself. No one can say who is responsible, or who should try to right the wrong, though everyone can see that wrong has been committed. It is the ultimate in collective responsibility.

Government will have to involve individuals if it is to be effective. The restrictions it imposes will foment discontent. State agencies are exasperatingly slow to change. Large, impersonal and sustained by a life of their own, they are invulnerable and inviolable except through the channels which they themselves have created, and thus control. The world they have brought about is an imperfect one. As a result, many people think that the only way to change the world embodied in them is to act outside their scheme of things: to cut red tape and resort to militant action. Young people are displaying a growing resentment against these institutions and the world they represent, and are tending to take the law into their own hands, demonstrating their disaffection more violently. If this disaffection is not to grow worse, a new method of government must be devised that involves ordinary individuals more than government does now. In future, government may come to exist in some form as yet unknown in either democratic West or centrally-planned East—a form probably somewhere in between, responsive to the opinions of individual citizens, through referenda conducted by computer-based opinion polls and by other means. If everyone can be made to feel involved, to feel that he or she has the power to shape the future, there is hope.

In his early exploitation of the environment, Man had to rely on natural materials and renewable organic resources. He used them without wondering much what their properties were, and why they were as they were. Growing dissatisfied with the rude and simple way of life, and using his natural inventiveness, he started to extract materials by physical and chemical processes, using them in more adventurous ways and in new, varied forms. He began to tap the accumulated deposits of fuel laid down during the course of the earth's history, and make other large-scale, irreversible changes in his environment. Technology's frontiers now extend to the very ingredients of life and the fundamental particles of matter. What of the future? Will Man try and press on? Or will he recoil, horrified at the way he has used his technological powers, and deliberately curtail technology's progress? Only the future will make that plain.

Suggestions for Further Reading

General

I Past

J. D. Bernal, *Science in History*, Watts and Co., London, 1954

R. A. Buchanan, *Technology and Social Progress*, Pergamon, Oxford, 1965

T. K. Derry and Trevor I. Williams, *A Short History of Technology*, Oxford University Press, London, paperback ed., 1970

Derek J. de Solla Price, *Science Since Babylon*, Yale University Press, Yale, 1961

Derek J. de Solla Price, *Little Science, Big Science*, Columbia University Press, New York, 1963

II Present

J. G. Crowther, *Discoveries and Inventions of the 20th Century*, Routledge and Kegan Paul, London, 1966

Peter F. Drucker, *The Age of Discontinuity*, Heinemann, London, 1969

Peter F. Drucker, *Technology, Management, and Society*, Heinemann, London, 1970

John Kenneth Galbraith, *The Affluent Society*, Pelican/Hamish Hamilton, London, 2nd ed., 1970

E. M. Hughes-Jones (ed.), *Economics and Technical Change*, Blackwell, Oxford, 1969

John Jewkes, David Sawers and Richard Stillerman, *The Sources of Invention*, Macmillan, London, 2nd ed., 1969

William R. Nelson (ed.), *The Politics of Science*, Oxford University Press, London, 1968

Simon Ramo, *Century of Mismatch*, David McKay, London, 1970

Theodore Roszak, *The Making of a Counter-Culture*, Faber, London, 1970

Technology and Economic Development, A Scientific American book, Pelican, Harmondsworth, 1965

Michael Shanks, *The Innovators*, Penguin, Harmondsworth, 1967

Daniel L. Spencer and Alexander Woroniak (ed.), *The Transfer of*

Technology to Developing Countries, Praeger (Pall Mall Books), London, 1968

Philip Sporn, *Technology, Engineering, and Economics*, MIT Press, London, 1969

Norman J. Vig, *Science and Technology in British Politics*, Pergamon, Oxford, 1968

III Future

Arthur Bronwell (ed.), *Science and Technology in the World of the Future*, Wiley-InterScience, 1970

Nigel Calder (ed.), *The World in 1984*, Penguin, Harmondsworth, 1965

Dennis Gabor, *Inventing the Future*, Secker and Warburg, London, 1963

Herman Kahn and Anthony J. Wiener, *The Year 2000*, Collier-Macmillan, London, 1967

Desmond King-Hele, *The End of the Twentieth Century*, Macmillan, London, 1970

More Specialized

I Electronics

S. Handel, *The Electronic Revolution*, Penguin, Harmondsworth, 1968

II Materials

D. W. F. Hardie and J. Davidson Pratt, *A History of the Modern British Chemical Industry*, Pergamon, Oxford, 1966

W. R. Jones, *Minerals in Industry*, Pelican, Harmondsworth, 4th ed., 1963

F. Sherwood Taylor, *A History of Industrial Chemistry*, Heinemann, London, 1957

Materials, A Scientific American book, W. H. Freeman and Co., London, 1967

III Computers

John Diebold, *Beyond Automation*, Praeger, New York, 1970

John Diebold, *Man and the Computer*, Praeger, New York, 1969

S. H. Hollingdale and G. C. Toothill, *Electronic Computers*, Penguin, Harmondsworth, 1965

Norbert Wiener, *Cybernetics*, MIT Press, Cambridge, Mass., 2nd ed., 1961

IV Engineering

Aubrey F. Burstall, *A History of Mechanical Engineering*, Faber, London, 1963

General Engineering Workshop Practice, Odhams Books, London, 3rd ed., 1963

V Energy

Margaret Gowing, *Britain and Atomic Energy 1939–45*, Macmillan, London, 1964

E. G. Sterland, *Energy into Power*, Aldus Books, London, 1967

VI Communications

Ronald Brown, *Telecommunications: the Booming Technology*, Aldus Books, London, 1969

Sidney Passman, *Scientific and Technological Communication*, Pergamon, Oxford, 1970

From Semaphore to Satellite, International Telecommunication Union, Geneva, 1965

VII Space

Philip Bono and Kenneth Gatland, *Frontiers of Space*, Blandford Press, London, 1969

P. J. Booker, G. C. Frewer and G. K. C. Pardoe, *Project Apollo: The Way to the Moon*, Chatto and Windus, London, 1969

Edwin P. Hoyt, *The Space Dealers*, John Day Co., New York, 1971

VIII Water

Robert Barton, *Oceanology Today: Man Exploits the Sea*, Aldus Books, London, 1970

J. L. Henson, *Water in the Chemical and Allied Industries*, Society of Chemical Industry, London, 1970

Philip Sporn, *Fresh Water from Saline Waters*, Pergamon, Oxford, 1966

William C. Walton, *The World of Water*, Weidenfeld and Nicolson, London, 1970

IX THE ENVIRONMENT

Robert Arvill, *Man and Environment*, Penguin, Harmondsworth, 1967

Arthur S. Boughey, *Man and the Environment*, Macmillan, New York, 1971

Rachel Carson, *Silent Spring*, Penguin, Harmondsworth, 1965

Paul R. Ehrlich and Anne H. Ehrlich, *Population, Resources, Environment*, Freeman, 1970

Richard H. Wagner, *Environment and Man*, W. W. Norton and Co., New York, 1971

A. N. Duckham and G. B. Masefield, *Farming Systems of the World*, Chatto and Windus, London, 1970

X DWELLINGS

Lewis Mumford, *The City in History*, Pelican, Harmondsworth, 1966

John Tetlow and Anthony Goss, *Homes, Towns, and Traffic*, Faber, London, 1965

Arnold Toynbee, *Cities on the Move*, Oxford University Press, London, 1970

XI MEDICINE

P. E. Baldry, *The Battle against Bacteria*, Cambridge University Press, Cambridge, 1965

Kenneth Cowan, *Implant and Transplant Surgery*, John Murray, London, 1971

Gerald Leach, *The Biocrats*, Cape, London, 1970

J. Rose (ed.), *Technological Injury*, Gordon and Breach, London, 1969

O. L. Wade, *Adverse Reactions to Drugs*, Heinemann Medical, London, 1970

XII WEAPONS

Nigel Calder (ed.), *Unless Peace Comes*, Penguin, London, 1970

Ronald W. Clark, *The Birth of the Bomb*, Phoenix House, London, 1961
Robin Clarke, *We All Fall Down*, Allen Lane, Penguin, London, 1968
Guy Hartcup, *The Challenge of War*, David and Charles, Newton Abbott, 1970
Richard D. McCarthy, *The Ultimate Folly*, Gollancz, London, 1970

Index